EDGE OF

Edge of Empire examines struggles over urban space in four contemporary First World cities: two sites in London and two sites in the Australian cities of Perth and Brisbane. Through these examples the spatialised cultural politics of a number of 'postcolonial' processes are unravelled: the imperial nostalgias of the one-time heart of empire, the City of London; the struggle of diasporic groups to make a homespace in the old imperial heartlands; the unsettling presence of Aboriginal claims for the sacred in the space of the modern city; and the emergence of hybrid spaces in the contemporary city. This book is about the spatial politics of race and nation, nature and culture, past and present. This is a 'global geography of the local'.

The book is distinctive in that it takes theories of colonialism and postcolonialism to the space of the city – it gives real space to the spatial metaphors of much contemporary social theory. If the contemporary city is a postmodern space it has not-so-hidden geographies of imperialism and postcolonialism. The global reach of the book – its focus on two poles of one trajectory of British imperialism – provides a global assemblage which forms a basis for understanding the unruly fortunes of imperialism over space and time. This is not simply a material geography of territory, it is also an imaginative geography of desire and memory.

Jane M. Jacobs is a Lecturer in the Department of Geography, University of Melbourne.

EDGE OF EMPIRE

Postcolonialism and the city

•

JANE M. JACOBS

Routledge

London and New York

First published 1996
by Routledge
11 New Fetter Lane, London EC4P 4EE

Simultaneously published in the USA and Canada
by Routledge
29 West 35th Street, New York, NY 10001

Routledge is an International Thomson Publishing company

© 1996 Jane M. Jacobs

Typeset in Garamond and Franklin Gothic by
Solidus (Bristol) Limited,
Printed and bound in Great Britain by
Redwood Books Ltd, Trowbridge, Wilts

All rights reserved. No part of this book may be reprinted or
reproduced or utilised in any form or by any electronic,
mechanical, or other means, now known or hereafter
invented, including photocopying and recording, or in any
information storage or retrieval system, without permission in
writing from the publishers.

British Library Cataloguing in Publication Data
A catalogue record for this book is available from the British Library

Library of Congress Cataloging in Publication Data
A catalogue record for this book has been requested

ISBN 0–415–12006–3 (hbk)
0–415–12007–1 (pbk)

CONTENTS

List of figures
viii
Preface
x
Acknowledgements
xii

1 TRAVELS ON THE EDGE OF EMPIRE
1
Real space
3
Itineraries
5
Talking out of place
7
The journey
10

2 (POST)COLONIAL SPACES
13
Colonialism and imperialism
15
Imperialism and space
19
The limits of the postcolonial
22
Postmodern space and the (post)colonial
29
Identity, the past and city space
34

CONTENTS

3 NEGOTIATING THE HEART
PLACE AND IDENTITY IN THE POSTIMPERIAL CITY
38

Difference gathered in the City of London
41

Making monuments
43

Picturing the empire
47

Pleasures of hearth
49

Imperial illusions
53

Continental entanglements
58

Colonial returns
64

4 EASTERN TRADING
DIASPORAS, DWELLING AND PLACE
70

Urban imperialisms
73

Land unoccupied
75

Hogarth and sag gosht
80

Developing nostalgias
87

Trading in community
91

Reinventing home
96

5 URBAN DREAMINGS
THE ABORIGINAL SACRED IN THE CITY
103

Ordering the urban
105

Visioning development
109
Urban nomadism
110
The erotic city
115
Brewery dreamings
116
Placing the Waugal
119
Back to Nature
121
Preserving the Crown
124
Fringedwelling
126

6 AUTHENTICALLY YOURS
DE-TOURING THE MAP
132
Nature, culture, colonialism
135
Imperial touring
138
Indigenous tourings
142
Remapping the colonial
149

7 CONCLUSION
157
Geographical encounters
157
Unruly imperialism
159
Postcolonial possibilities
161

Bibliography
164
Index
187

FIGURES

•

1	Neils M. Lund's 1904 painting, *The Heart of the Empire*.	39
2	The modernist Mies van der Rohe scheme for Bank Junction.	42
3	The James Stirling and Michael Wilford scheme for Bank Junction.	42
4 and 5	The nineteenth-century buildings on the Bank Junction development site: before and after.	44–45
6	Map of Bank Junction showing listed buildings and Conservation Areas.	46
7	The glimpsed view of St Paul's from Cornhill, before and after development.	50
8	Protected views of the dome of St Paul's Cathedral.	52
9	James Stirling's Neue Staatsgalerie in Stuttgart.	62
10	Traffic regulation responses to the IRA bombings.	66
11	The City of London's 'Ring of Steel'.	66
12	The Georgian housing stock of Spitalfields.	76
13	Map of listed buildings and Conservation Areas surrounding Spitalfields Market.	77
14	The Spitalfields Trust 'flying squad'.	78
15	The Delft fire surround of the Spitalfields community.	82
16	Spitalfields according to the developer's history.	84
17	The too-Georgian vision for Spitalfields.	86
18	The Spitalfields Development Group logo.	88
19	The MacCormac scheme for the Spitalfields Market redevelopment.	89
20	Signs of racial harassment in Spitalfields.	92
21	Anti-racism street march in Spitalfields.	93
22	The Campaign to Save Spitalfields cartoon.	94
23	The Old Swan Brewery.	104
24	The gridded city of early Perth.	106
25	Heritage sign at the Old Swan Brewery.	117

26　The Roger Gregson plans for the Old Swan Brewery
　　redevelopment. 123
27　Robert Bropho (*left*) and a fellow protester at the Old Swan
　　Brewery. 128
28　Police remove protestors from the Old Swan Brewery site. 129
29　The Brisbane City Council ecocentre. 133
30　The panoptic view offered to turn-of-the-century sightseers at
　　Mount Coot-tha. 141
31　The painted images used at the J. C. Slaughter Falls art trail. 144
32　Artists Laurie Nilsen and Marshall Bell of Campfire Consultancy,
　　the 'Main Gallery' at J. C. Slaughter Falls. 145
33　Re-mappings at the J. C. Slaughter Falls art trail. 150
34　Laurie Nilsen's Aboriginalised re-mapping of the city of Brisbane. 152

PREFACE

•

Studies in colonial discourse analysis and postcolonialism have radically altered the way imperialism, the nation and race are understood. Recent colonial and postcolonial theory is saturated with a spatialised language although more often than not it functions metaphorically. This book seeks to move beyond the spatial rhetoric of colonial and postcolonial theory and return it to 'real' geographies. I am concerned with showing how the materiality of place, the imaginative spatialities of desire and a cultural politics of territory are fundamental parts of colonial and postcolonial formations in the present. I pursue this project through the example of contemporary conflicts over space in cities in Britain and Australia.

This book brings together four sites in cities which were generated under, or energised by, the British imperial project. Two of these sites are located in London, once the heart of empire, and two are based in cities in Australia, once the geographical edge of that empire. The empirical range of the book says much of my own positioning, as an 'anglo-Australian', with a very particular experience of colonialism. There is as much ambiguity as certainty in this positioning. As an *anglo*-Australian I am marked in my country of birth as a member of the culture of colonisation, a status which problematises my engagement with issues relating to the politics of Aboriginal cultural sites. As an anglo-*Australian* in Britain, I am 'of the colonies' and when researching in London I was, at times, received merely as a more earnest version of the Australian backpacker. I am both coloniser and antipodean, a representative of the European core but also other to it (McLean 1993: 18). No doubt my position as someone born in a British settler colony, in a country now agonising its way towards being a republic, shaped the way I viewed the imperial nostalgias and racist practices of the London of the 1980s. Without doubt, my ability to discuss contemporary Aboriginal interests in land and Australian racism is constrained and directed by my own lineage as a daughter of the colonising generations.

The title *Edge of Empire* does not refer to sites literally on the periphery of an imperial geopolitical regime. The British empire has in all but a few cases officially ended and the nations and peoples once part of its reach are in a state precariously registered as postcolonial. There is in theory no empire, no centre, no edge. Of course this is an official state which is challenged by the present as we know it: there are centres, there are peripheries, there are persistent structures of domination and subordination. The ongoing presence of anti-colonial politics attests to this. But the social and spatial demarcation of such uneven politics is no longer as clear as it once was. The structures of power that gave rise to empire live on in a more disorganised fashion. And any lingering certainty imperialism has is daily challenged by diasporic settlements, new nationalisms, indigenous land rights claims as well as a plethora of other events. That is, British imperialism lives on in the present but it is also always at its 'edge' point. This book describes this politics of the 'edge' in terms of the unstable negotiation of identity and power which occurs in and through the space of the contemporary city. The 'edge' of the title *Edge of Empire* evokes not a literal edge, the periphery, but what bell hooks (1991: 149) describes as a 'profound edge', the 'unsafe' margin which marks not only a space of openness but also the very negotiation of space itself.

ACKNOWLEDGEMENTS

•

This book is a product of a reappraisal of disparate research I have been involved in since the early 1980s and it would be impossible to acknowledge all those who have helped me to formulate my ideas and given me access to information. Perhaps my greatest debt is to those who I see as having first forged my sense of the politics of race and place. There is nothing in this book about the remote South Australian community of Port Augusta but it was there as a most naive researcher under an uncompromising Aboriginal tutelage that I painfully learnt the lessons of racism and colonialism, saw the contingency of postcolonialism and began to understand the politics of identity and place. The lessons of Port Augusta have traversed the globe with me.

I would also like to thank the various organisations and individuals who gave of their time and their records. In particular I would like to thank the Campaign to Save Spitalfields from the Developer and especially Jil Cove, the Spitalfields Trust, SAVE Britain's Heritage, the CARE Campaign and especially Reverend Victor Stock, Brisbane City Council and Marshall Bell of Campfire Consultancy. Thanks are also due to the various individuals and organisations, acknowledged in the text, who gave permission to reproduce figures. Chapter 3 of this book is a version of an article which appeared in *Environment and Planning D: Society and Space*, and I acknowledge Pion Publishing for allowing this article to be reproduced, albeit in a substantially altered form.

Books also have their practical life. Time and financial support for the writing of this book was provided by the Australian Research Council Postdoctoral Fellowship scheme. Space for thinking and writing and technical support was provided by the Departments of Geography at both University College London and the University of Melbourne. Special thanks go to Chandra Jayasurya, Wendy Nicol, Natalie Jamieson, Mary Quilty, Tim Aspen and staff and Chris Cromarty who, in various ways, assisted in the production of the text. The Centre for Research in Women's Studies and Gender Relations at the University of British Columbia, the Department of Geography at the

University of Sheffield and the homes of Paula, Jo and Peter provided havens during the final haul. Various colleagues have listened, guided and inspired and special thanks are also due to Kay Anderson, Jacqueline Burgess, Dipesh Chakrabarty, Felix Driver, Fay Gale, Ken Gelder, Derek Gregory, Peter Jackson, Michael Keith, Sylvia Kleinert, Fiona Nicholl and Charlotte Townsend-Gault.

In my life family and friends have luxuriously blended and I would like to dedicate this book to what I consider to be my extended family. First, to Hannah, Ken and Christian who have provided friendship and humour. And, of course, to my mother and father, Barbara and Allen Jacobs, who not only gave me their enduring support but also bestowed upon me the name 'Jane' which, in my academic life, has taught me well the play of 'otherness' and the unpredictable fortunes of signification.

1
TRAVELS ON THE EDGE OF EMPIRE

•

Every voyage can be said to involve a re-siting of boundaries ... an undetermined journeying practice.

(Minh-ha 1994: 9)

In April of 1993, I undertook a peculiarly academic journey which took in both a conference and a field trip. From my home in Melbourne, I travelled to Sydney to attend a conference on postmodern cities. From there, I flew to central Queensland, where I was met by an Aboriginal historian, and trekked across dirt roads to visit her mother's country and to start a project of Aboriginal and non-Aboriginal joint writing about place.[1] Within the space of two days I had gone from discussing 'postmodern' cities to participating in self-consciously 'postcolonial' fieldwork. In Sydney, as I listened to various papers on postmodern cities, I could not help but wonder why the processes and formations being discussed were infrequently connected to colonialism, imperialism or postcolonialism. As I walked through the bush of central Queensland, I wondered why the 'postcolonial' politics of identity and place which generated this field trip was often only incidentally traced in analyses of the city. This book is an attempt to forge a productive encounter between the space of the contemporary city and recent theorisations of colonialism and postcolonialism, which is achieved through detailed accounts of the cultural politics of identity and place in four contemporary urban settings in Britain and Australia.

The relations of power and difference established through nineteenth-century British imperialism linger on and are frequently reactivated in many contemporary First World cities. Yet in these cities there are also various challenges made to imperialism by way of what might be thought of as postcolonial formations. These expressions and negotiations of imperialism do not just occur *in* space. This is a politics of identity and power that articulates itself *through* space and is, fundamentally, *about* space. This is plain enough to see in terms of the grand territorialisations of the building of empire, but it is also evident in the politics associated with contemporary processes of urban redevelopment, which is the main focus of my book. Changes to city space may occur within a calm consensus but more often than not they result in protracted struggle or in self-conscious gestures of reconciliation. This politics is rarely only about how space is to look and function, about competing architectural aesthetics or urban planning ideologies, although such concerns may well provide the dominant discursive form of these struggles. These place-based struggles are also arenas in which various coalitions express their sense of self and their desires for the spaces which constitute their 'home' – be it the local neighbourhood or the nation home, an indigenous home or one recently adopted. The politics produced by places in the process of becoming or being made anew is, then, also a politics of identity in which ideas of race, class, community and gender are formed. This politics of identity and place is not simply built around structures of power internal to the city itself or even to globally linked processes of urbanisation. It is undeniably a politics that occurs in and is concerned with the city, but for many groups it is also a politics constituted by a broader history and geography of colonial inheritances, imperialist presents and postcolonial possibilities.

Although imperialism is undeniably a political and economic event, it also operates through a range of cultural processes. For example, social constructs of Self and Other provided the fundamental building blocks for the hierarchies of power which produced empires and the uneven relations among their citizenry. Under colonialism, negative constructions of the colonised Other established certain structures of domination through which the coloniser triumphed. Similarly, counter-colonial challenges frequently involve subordinated groups reclaiming 'precolonial' identities or revalorising identities 'made' under the force of colonialism. The processes by which notions of the Self and Other are defined, articulated and negotiated are a crucial part of what might be thought of as the cultural dimension of colonialism and postcolonialism. In the first instance, these processes mark out the very categories of difference which have come to be the positively or negatively ascribed 'cultures', 'races', 'ethnicities' or 'genders' of imperial structures of power. But also, the very making and remaking of identity occurs through representational and discursive spheres, both official and popular, material and ideological. As Thomas (1994: 2) notes, colonialism has always been 'imagined and energised through signs, metaphors and narratives'.

The nineteenth-century imperial project most clearly, but not exclusively, depended upon racialised notions of Self and Other. Imperialism operated within an ideal of the Manichaean binary, which constructed a demonised Other against which flattering, and legitimating, images of the metropolitan Self were defined. Such racialised constructs were never stable and were always threatened not only by the unpredictability of the Other but also the uncertain homogeneity and boundedness of the Self. Indeed, the vitality of such binary constructs is most likely a result of their being anxiously reinscribed in the face of their contested or uncontainable certainty. It is, in part, this anxious vitality that gives racialised categorisations elaborated under colonialism such a long life and allows them to remain cogent features even of those contemporary societies that are formally 'beyond' colonialism.

Colonial discourse and power has operated through a complex intersection of social constructs based around race, gender, class and sexuality (see Ware 1992; C. Hall 1993). The emphasis of my book is on a racialised politics of differentiation although this is not intended to relegate other constructs to the sidelines or to say that feminist theory has nothing to offer the rethinking of imperialism. For example, feminism has shown the intimate link between imperialism and what Spivak (1988a: 107) calls 'the practice of masculism': the way in which imperialism depended upon masculinist possession of 'virgin' lands and patriarchal tamings of feminised wildness (Shohat 1991). But also, it is feminist theory which has offered some of the most radical rethinking of the politics of identity.

REAL SPACE

In recent colonial discourse analysis and postcolonial theory spatial metaphors proliferate. Here the contours, boundaries and geographies of space are 'called upon to stand in for all the contested realms of identity' (Vidler 1992: 167). Smith and Godlewska (1994: 7) suggest that many of the literary-based re-evaluations of colonialism are ambivalent about geographies 'more physical than imagined'. The possibility of material and imagined geographies being neatly separate, as Smith and Godlewska imply in their critique, is of course unthinkable – one constitutes the other. Yet it is true that in recent social theory the spatial is metaphorically everywhere but oft-times nowhere (Massey 1992; Smith and Katz 1993). It is in this context that my book attends to what I somewhat unfashionably refer to as the 'real' geographies of colonialism and postcolonialism.

Edward Said (1989: 218) has reminded us that empires could not have been 'without important philosophical and imaginative processes at work in the production ... acquisition, subordination and settlement of space'. For example, the role of cartographic constructions of space in the building of empires has been well documented by geographers (see, as examples, Harley 1988, 1992; Livingstone 1992). These imaginative geographies of Reason regulated, bounded and secured space as a precondition for the embodied occupations which followed and the subsequent incorporation of these territories into the global power grid of empire. Because of the primacy of the spatial in imperial projects, postcolonial politics is also often explicitly spatial (Said 1993: 271). At the most general level this can be seen in the way nationalist struggles and non-sovereign land rights claims often focus on regaining control of territories lost to the spatial drive of imperial expansion.

The city is also an important component in the spatiality of imperialism. It was in outpost cities that the spatial order of imperial imaginings was rapidly and deftly realised. And it was through these cities that the resources of colonised lands were harnessed and reconnected to cities in imperial heartlands. Under the explanatory reach of world-system theories of capitalism, these cities later came to be understood as part of an interdependent 'First World' and 'Third World'. The concern of this book is not with the colonial city as a thing of the past or even with the contemporary Third (or First) World city as an outcome of global capitalism *per se*. My concern is with the way in which the cultural politics of place and identity in contemporary First World cities is enmeshed in the legacies of imperialist ideologies and practices. It is true that many of the urban transformations I examine – gentrification, mega-developments, heritage designations and spectacles of consumption – are widely understood as hallmarks of postmodernity. If such urban change does mark a postmodernity, it is a postmodernity with not-so-hidden histories and not-so-absent geographies of imperialism.

Roland Barthes (1981: 96) reminds us that the city is the very 'place of our meeting with the other'. In contemporary cities people connected by imperial histories are thrust together in assemblages barely predicted, and often guarded against, during the inaugural phases of colonialism. Often enough this is a meeting not simply augmented by imperialism but still regulated by its constructs of difference and privilege. This is most clearly shown in the racial politics of cities and especially in the often startlingly spatial outcomes of segregation or inner-city racialisation. Not unrelatedly, imperialism may also be reactivated in the present through various nostalgias which seek to memorialise the period of imperial might. Such trends may be marked by the self-conscious elaboration of tradition, or in the preservation of historic buildings or through the emergence of new industries of consumption which build on the past, the primitive, Nature. Imperialist manipulations of space never had an unchallenged surety, either in the past or the present. Precisely because cities are sites of 'meetings', they are also places which are saturated with possibilities for the destabilisation of imperial arrangements. This may manifest through stark anticolonial activities, but also through the negotiations of identity and place which arise through diasporic settlements and hybrid cultural forms.

The politics of identity and difference established under colonialism and negotiated through a range of postcolonial formations is not only 'practiced' in particular settings, as if they are simply 'staging grounds' (Appadurai 1990: 15), but also activated through 'real' space. This is not to suggest an abstracted space with determining force, nor a material space which is outside of social relations. Rather, it is to propose, as Doreen Massey (1994: 3) does, that space is a part of 'an ever-shifting social geometry of power and signification' in which the material and the ideological are co-constitutive. This is by no means a settled notion of space, but rather a troubled social/spatial dynamic. It is, to toy with one of Said's (1993: 6) notions, a 'geography which struggles'. These spatial struggles are not simply about control of territory articulated through the clear binaries of colonialist constructs. They are formed out of the cohabitation of variously empowered people and the meanings they ascribed to localities and places. They are constituted from the way in which the global and the local always already inhabit one another. They are products of the disparate and contradictory geographies of identification produced under modernity. These struggles produce promiscuous geographies of dwelling in place in which the categories of Self and Other, here and there, past and present, constantly solicit one another.

ITINERARIES

This book is structured around detailed readings of four sites of transformation in contemporary, First World cities. For the most part my concern is with the present (defined as the 1980s to the early 1990s) although clearly the past and ideas about the past come into and constitute this present. The four cases include two sites in the old imperial heartland of London, the City of London and Spitalfields, and two sites in cities once thought of as on the edge of the British empire, the Australian cities of Perth and Brisbane. These sites share an experience of British imperialism. In assembling this itinerary of urban sites I do not construct a strict model of cause and effect, rather, the basis of a loosely comparative project (Clifford 1992: 105). This is not an account of the neat linear flow of Englishness to Australia, or the faintly traced counterflows of indigenous Australia to the imperial heart. In placing these sites together I seek to show how the imperial project is both global in scale but also messy in its local effects.

It is undeniable that my selection of these Australian and British urban sites was directed less by a measured process of 'sample' than by my own idiosyncratic placement in a particular colonial history.[2] Yet this accident suited my intent. My aim was not to assemble a First World city, which signifies a 'transcontinentally active core', and a Third World city, which might stand for a 'locally bounded periphery', and compare (Lowenhaupt Tsing 1994: 282). My assemblage of city sites establishes an empirical framework in which notions of core and periphery are, from the outset, destabilised. First, the sites in Australian cities register the First World or the 'west' as it exists outside the over-privileged geography of North America and Europe. But the Australian examples also show that there is an often forgotten Fourth World – Aboriginal Australia – within such First World cities. Second, the sites in London show how a First World city, precariously reclassified as 'global', is also a diasporic city which contains a Third World within its boundaries. These four urban sites each establish exceptions to the core/periphery rules which have shaped the way in which the international arrangements of power and privilege generated by imperialism have been conventionally registered.

My loyalty to the integrity of four specific cases studies may well be an example of the recalcitrant but predictable empiricism of the geographer. But this loyalty is based on the belief that it is through the local, rendered in detail, that the complex variability of the (post)colonial politics of identity and place can be known. These detailed accounts of the local mark out the 'tendential lines of force' by which imperialism holds, and is challenged, in the contemporary city (Frankenberg and Mani 1993: 307). In considering the local as I do, I am not proposing a retreat to nostalgic localism or a spatialised relativism. The translocal/transnational tendencies of colonialism and, more latterly, global capitalism, have not resulted in the obliteration of the local or the diminution of the national. The 'logic of capital' in both its imperial and presumably postmodern forms has operated through difference, through the specificity of the local. To focus on the local is also, then, to attend to the global (Massey 1993a).

In building my argument through four specific urban sites this book unavoidably takes the form of a journey and is indeed a journey under the guidance of a geographer. I admit my disciplinary positioning, well aware of the role the pragmatic geographies of description and cartography have played in imperialism and its colonial effects (Driver 1992). These geographies sought to fix and give transparent sense to the unknown. Unlike the geographical practices of imperialism my aim is not to demarcate the exotic or to settle space in a masculinist gesture of creating a single vantage point (Bordo 1986; Hartsock 1990). The urban 'visits' made in this book seek to uncover the unsettled geography of space and place and the shifting boundaries of identity within the contemporary moment. I do not want to make these sites transparently readable, but rather to approach them in the spirit of the figure of the 'urban-culturalist', whose task is to show how the familiar is often already strange (Grossberg 1988: 378, after Morris 1988a).

Travel metaphors are fashionable (see, as examples, Chambers 1994a and Robertson *et al.* 1994) and the tendency in current critical thought to evoke such metaphors might be, as Chambers (1994a: 3) suggests, 'facile'. Feminist theorists, in particular, have expressed concern about the notion of the 'travelling theorist', not because it is limp but because it may be too potent. Morris (1988a) points to the way in which masculinist ways of knowing and patriarchal relations of power are complexly entwined in the practice of travel and the writing of travel stories (see also Blunt and Rose 1994). This is not simply an issue of the differential access to travelling that women historically have had. Travel stories written as 'Voyages and Maps', Morris suggests, 'relentlessly generate models of the proper use of place and time – where to begin, where to go, what to become in between' (1988a: 35). Others have argued that the travel metaphor may well be *too* nomadic to accommodate, say, the situated oppressions of women of colour and the specifically located theory/politics of racialised minorities and women (hooks 1992; Wolff 1993). For many minorities, travel is ambivalently placed: something imposed, something from which they are excluded, something that brings them into contact with the 'terrorizing force' of supremacy (hooks 1992: 344).

Chambers (1994a: 3) suggests that the turn to 'travel' may simply be a theoretical fad which, in a most imperialist mode, attempts to keep hold of a proliferating diversity. But in many respects it is precisely the nomadic qualities of travel that have worked to secure its place as a way of thinking and doing contemporary critical analysis. Under conditions that are as much about accelerated processes of globalisation as they are an intensification of a situated politics of difference, traditional ways of knowing are registered as saying too much from a now all-too-fixed vantage point. Theory sensitised to movement and studies that are configured, as this one is, around diverse assemblages of people and place, are not simply seductive but, I would argue, necessary. Trinh T. Minh-ha (1990: 334 and 1991: 188), for example, suggests that 'mobility' is precisely what is needed to displace the arrangements of power which have oppressed all women as well as people of colour. Such arguments suggest that 'travelling theories' can be productively redeployed. They can mark an honest entry point into the contradictory condition of a present that is as much about the global as it is the local, as much about identity and place as it is a more fractured 'politics of location' (hooks 1992: 343).

TALKING OUT OF PLACE[3]

The relatively recent period of self-reflexivity within the new human sciences has combined with postcolonial claims for rights over knowledge and generated new protocols of speaking and writing. Although ensuring what are absolutely necessary and productive adjustments to the terms of authority, the combined effect has been to exaggerate the distinction between analyses of colonial discourses (which are safely available for all to see and read) and analyses of the experiences and formations of those Othered by colonialism (which are seen and read by those 'outside' at their own peril). This book self-consciously transgresses this now unproductively amplified distinction. In the chapters that follow I do not confine my analysis to colonialist constructs, but dare to translate and interpret the ways in which those demarcated as Other negotiate metropolitan structures of power. That is, I also take the risk of presenting and interpreting the views and positionings of those marked as Other in the imperial imagination. I do this fully aware that such intercultural interpretation is never 'politically innocent' and that I cannot, and do not, divest myself of my position within a not-so-fraying imperialist world (Clifford 1992: 97). But I take this risk with the conviction that to confine my attention to the workings of colonialist power, without consideration of how colonialism encounters and is transformed by those it seeks to dominate, I cannot possibly claim an anticolonial politics and, what is more, might simply work to embellish the 'core'. I might invoke a postcolonial notion of some 'in-between space' as a way of legitimating my transgression. I might argue that postcolonial negotiations are 'spaces' that are conveniently not of the Centre nor of the Other, an orphaned surplus of hybridity for which anyone might speak. Such 'spaces' are produced in the politics of colonialism and postcolonialism. Yet while the implications of hybridity for the issues of authorisation are ambiguous, this should not establish opportunities for speaking which are outside an unavoidable politics of power. Responding to this politics does not simply mean I cannot speak because I am not 'from the right place' or I can now speak too much because our respective places are unbound (Spivak and Young 1991: 228). It means that one's speaking must be measured by a responsibility to an anticolonial politics.

The accounts presented of these sites and their cultural politics of production are not univocal. To avoid univocality is not simply to say many people see one place in different ways or to establish a now more conversant binary, as Said does in his notion of an 'atonal ... contrapuntal' interplay of Self and Other (1993: 59–60). Feminist theorisations of subjectivity in particular have worked to 'trouble' fixed notions of identity and difference (Butler 1990a, 1990b; Ferguson 1993: 154). This more 'mobile' approach to the nature of identity and difference can be used productively in considering the relationship between identity and place. It unsettles the notion of a bounded, pre-given essence of place to which the identity of those who dwell there adheres. It attends instead to the constant interplay between positional and variable metropolitan histories and other histories and the complex intermeshing of the global and the local. That is, it attends to what Said (1993: 56) refers to as the 'overlapping territories' and 'intertwined histories' which produce the unstable conditions of dwelling in place. Within this politics essentialised notions of identity do not disappear, but they are understood as positional constructions framed within certain arrangements of power. This book uses the local to show the adaptive persistence of imperial structures of power, the always present postcolonial counterflows and the unanticipated trajectories of identity and power produced within this negotiable politics of difference.

Much of the analysis focuses on institutional discourses about urban space: city plans, conservation reports, parliamentary statements and planning inquiry transcripts. I also draw upon an array of statements and images produced by interest groups actively participating in the political struggles around urban development and change. My analysis reaches out beyond the narrow discursive and representational spheres that were produced by these struggles over urban space and incorporates a broader political and economic context. I do this not to propose that the discursive and symbolic fields which are associated with urban redevelopment are simply subsidiary effects of such contexts. Rather, I show how these discursive and representational practices are in a mutually constitutive relationship with political and economic forces. Together, they actively create the material and imaginary landscapes of the city.

This book moves self-consciously towards a cultural politics of place as opposed to a reading of a textualised landscape (see Duncan 1990, Barnes and Duncan 1992; Duncan and Ley 1993). I make this distinction explicit because of a concern with the problematic of a textual conceptualisation of space both in terms of the intersection of identity and place and the obligations of a postcolonial political agenda. In their most narrow conceptualisation, textualised readings over-privilege the built form and the visioned urban plan, which are themselves a mark of power, 'a material manifestation of dominant interests' (Gottdiener 1986: 214–215). The material artefacts, the built forms, of cities are in many senses the displaced 'main attractions' of my urban journeyings. I am more concerned with the complicated politics of the production of urban space, than the object produced. Struggles about how an urban space is to be used and how it is to look often go on for years – capturing public resources, mobilising disparate groups into political action and generating one vision after the other. This protracted politics of production is in itself a social and material formation which has effects which not only precede but reach well beyond the space under contest. Often then the 'exemplary object' of these studies is not that which *is* but that which is *not yet*. In such contexts it is not the 'object' which is *the* thing to be 'read' but 'change itself' (Morris 1990: 12). More broadly, the claim to readability in 'textual' geographies resonates uncomfortably with the transparency assumed by imperial visionings (de Certeau 1984: 92–93; and see Demeritt 1994 and Gregory 1994: 146–147). The geographical articulations of imperialism are not simply laid out across the landscape for (yet another) reading. They exist in the 'opaque' intersections between representational practices, the built form and a range of other axes of power which determine the precise historical context within which imperialism does and does not hold. This includes the uneven geography of capital investment, legal and judicial regulatory regimes as well as the various territorialisations and deterritorialisations of space which occur through protest, violence, ironic artistry, or simply dwelling in place.

THE JOURNEY

The main part of this book provides detailed accounts of specific sites of urban redevelopment and the politics of identity articulated through them. Although each chapter develops quite specific theoretical points, I begin the volume with a more general discussion which outlines my approach to colonialism, imperialism and postcolonialism. I do this with specific reference to the cultural politics of identity and place in the urban context, thereby establishing the foundations upon which the detailed case studies are built.

The spatial imaginary of British imperialism was based on the forceful flow of power from what came to be understood as the 'core' to what was relationally designated as the 'periphery'. In the text the case studies are arranged in a manner which mirrors the original spatiality of British imperialism. The first case studies are taken from London, the former heart of the British empire. The last two studies are taken from Australian cities, a former edge of that empire. My tracing of the original spatial logic of one trajectory of British imperialism is not intended to re-inscribe the logic of core and periphery. It is an ironic device which amplifies the precarious nature of this logic in the setting of the contemporary city.

Each of the case-based chapters marks distinct ways in which the cultural politics of colonialism and postcolonialism is articulated in the present. In the chapter on the City of London (Chapter 3), I tackle the interplay between a recently energised 'global' city and the idea of empire as expressed in a long-running planning struggle around the redevelopment of the historic built environment. In the second of the London case studies I turn to the inner East London neighbourhood of Spitalfields where a large Bengali British population now resides (Chapter 4). In this chapter I trace the way in which neighbourhood change, in the form of gentrification and mega-scale redevelopment, produces a tension-filled encounter between emergent desires for a 'multicultural' Britain and the local Bengali community. In the first of the Australian examples, drawn from the city of Perth, I examine the way in which colonial urban development repressed Aboriginal interests in land and examine the uncanny reappearance of the Aboriginal sacred in the secularised space of the city (Chapter 5). And in the final case study, I deal with the cultural politics of new tourisms in the city of Brisbane and in particular the way the city is actively reinventing itself through Nature and Aboriginality (Chapter 6).

The case studies stand somewhat independently of each other but none the less connect and unavoidably rub against one another. For example, the emergence of a 'global' City of London and the local changes this generates (Chapter 3) establishes part of the context for the redevelopment struggles in Spitalfields (Chapter 4). The London chapters also provide important contextual parameters for the Australian chapters. For example, the political and economic restructurings of the 1980s, encountered first through the London material, were registered in cities all over the world, including cities in Australia. Similarly, the urban heritage sensibilities evident in the London cases are part of a much broader planning common sense which also operates in Australian cities. The urban planning controversy in Perth (Chapter 5) was, in part, shaped by these global economic and cultural formations. Finally, some of the historical material on Aborigines under colonialism (Chapter 5) establishes contextual points for the exploration of the racial politics of new industries of consumption (Chapter 6).

Together these chapters work to show the diversity and adaptive realignments of imperialism, but also the complex range of postcolonial formations which unsettle, negotiate and at times overtly resist imperialist structures of power. The places visited are not static sites but sites in the process of becoming. They are places saturated with the politics of transformation: be it expressed as conflict, through calm consensus or in self-conscious gestures of reconciliation. Together these places mark a geography in which centre and margin, Self and Other, here and there are in anxious negotiation – where there is displacement, interaction and contest (Clifford 1992: 101). I have contrived this 'global' geography of the 'local' with the conviction that it will uncover tensions and disjunctions which are the very space of a productive critique of the cultural politics of imperial power in the present.

NOTES

1. This joint project was with Jackie Huggins and Rita Huggins and was based in the Carnarvon Gorge area of central Queensland (see Huggins, Huggins and Jacobs 1995).
2. I discuss my positioning as an anglo-Australian in the Preface to this book.
3. This sub-heading is taken from an article jointly written with Ken Gelder (Gelder and Jacobs 1995a).

2

(POST)COLONIAL SPACES

•

Difference is encountered in the adjoining neighborhoods, the familiar turns up at the ends of the earth.

(Clifford 1988: 14)

In recent years imperialism and postcolonialism have attracted unprecedented academic attention. The emphasis of much of this new work is decidedly cultural, emerging as it does from literary studies, but its effect has reached into a wide range of disciplinary fields. There is little doubt that Edward Said's *Orientalism* (1978) established a template for studies alert to the 'culture' of imperialism. As Williams and Chrisman (1993: 5) suggest, Said's approach 'inaugurated' the field of study which has come to be known as colonial discourse analysis. Such studies show the ways in which discursive formations worked to create a complex field of values, meanings and practices through which the European Self is positioned as superior and non-Europeans are placed as an inferior, but necessary, Other to the constitution of that Self. Such metropolitan constructs of Self and Other were integral to the territorial, military, political and economic extensions of European power across the globe, the processes known as colonialism and imperialism. As Said (1995: 332) emphasises, processes of social construction of identity are not simply 'mental exercises', but also 'urgent social contests involving ... concrete political issues' such as territory, violence, law and policy. Social constructs and the meanings and practices they generate are at the very heart of the uneven material and political terrains of imperial worlds.

As the work on the nexus of power and identity within the imperial process has been elaborated, so many of the conceptual binaries that were seen as fundamental to its architecture of power have been problematised. Binary couplets like core/periphery, inside/outside, Self/Other, First World/Third World, North/South have given way to tropes such as hybridity, diaspora, creolisation, transculturation, border. James Clifford (1994: 303) refers to this proliferation of a new analytic language as an 'unruly crowd of ... terms'. This new language is associated with what has come to be known as a postcolonial perspective and owes much to those writing from and about 'the margins' and the vigilantly deconstructive potential of poststructuralism. Such terms provide the conceptual frames for interpreting a range of new (or newly found) phenomena which might be loosely described as counteracting or unsettling imperialism. But they also offer the means for rethinking familiar imperialist structures of domination whose complexities were once obscured by neat binary classifications. The postcolonial critique has mobilised a new conceptual framework which, while not denying the efficacy of imperialist structures of domination, uncovers their often anxious contingency and internal variability. This is not to deny that binary notions of Self/Other did not inhabit the imperial imagination, but rather to show that this was an intensely unstable arrangement in which these notions of Self and Other, as Derrida (1982) puts it, always 'solicited' each other and produced decidedly disruptive effects (Gelder and Jacobs 1995a: 153).

What is common to these perspectives is an attention not to the formal geopolitics of imperialism nor to a singular political economy of 'the world', but a sensitivity to the culture of imperialism and those formations that might register as counter-colonial or postcolonial. For example, colonial discourse studies attend, as Thomas (1994: 42–43) suggests, to the meanings and values that energise colonialism. The value of new perspectives on imperialism and postcolonialism moves beyond seeing how culture 'becomes' governmentality through the pragmatic deployment of social constructs (ibid.). One of the key contributions to be made by postcolonial studies is to demonstrate the vulnerability of imperialist and colonialist power. Postcolonial studies highlight the way these cultures of power and domination never fully realise themselves. They are always anxiously regrouping, reinventing, and reinscribing their authority against the challenge of anticolonial formations but also against their own internal instability. This is not to say that imperialism does not take hold and create most emphatically material and painfully uneven geographies. Rather, it is to reach into the ambivalent cultural politics of this domination so that the necessity of its tenacious and adaptive power can be better understood. It is also to register the various ways in which colonised groups subvert this power not simply through stark opposition but also through disruptive inhabitations of colonialist constructs.

While it is enticing to think of the present as somehow already postcolonial, this 'postcoloniality', as the term itself suggests, is still deeply entwined with colonial formations. Colonial constructs not only belong to a past that is being worked against in the present, but also to a past that is being nostalgically reworked and inventively adapted in the present. Just as postcolonialist tendencies have always been produced by colonialism, so colonialist tendencies necessarily inhabit often optimistically designated postcolonial formations. I do not want to suggest a grim entrapment in colonial formations, past and present. The cultural and spatial processes I describe in this book point to the anxious tenacity of colonialist tendencies. The 'postcolonialisms' described hereafter are not always neatly 'against' colonialism's residual and revived formations, part of the seductive realm of resistance. I do not deny the possibility of resistance but instead I suggest that it is one articulation of many which work against or slip outside of colonialism. The colonised engage not only in resistance but also in complicity, conciliation, even blithe disregard. It is a revisionary form of imperialist nostalgia that defines the colonised as always engaged in conscious work against the 'core'.

This proliferation of new ways of being and seeing in the field of imperial and postcolonial studies has generated its own confusions. What is the difference between colonialism and imperialism? What is postcolonialism, how real is the social condition it presumes to describe, and what are the limits of this term? Does the emphasis on postcolonialism displace or productively elaborate the politics of race? How does attention to the cultural sphere work to destabilise the surer spaces of power and difference marked by geopolitical or political economy perspectives? What has been the role of space in colonial and postcolonial projects and how might we rethink the space of the city in these terms? And how can the spatial discipline of geography move from its historical positioning of colonial complicity towards productively postcolonial spatial narratives? In exploring these questions I draw on examples which are of direct relevance to, and thereby help give context to, the local studies that follow. In particular, my attention is confined to the case of the British imperial project and Australian colonisation and to the specific role of the city in colonial enterprises and as a site of postcolonial formations.

COLONIALISM AND IMPERIALISM

Once, when talking about contemporary London to a British audience, I chose to categorise the now of London as 'postcolonial'. It was suggested by a member of the audience that what I really meant was 'postimperial'. We agreed it was post, but post what? What was colonialism to me was imperialism to him. I recall this interchange because it demonstrates how imperialism and colonialism are terms that are, in part, differentiated by positioning within the power geography of empire itself. The exchange designated imperialism as 'belonging' to the metropolitan core: addressing its cultural constructs and its accumulative drives and territorial expansions. Nineteenth-century British territorial expansion, in this corrective exchange, was registered as imperialism at the core but colonialism at the edge or, to put it differently, imperial in intent, colonial in effect.[1]

Confusions such as these commonly surround the terms colonialism and imperialism and what may be registered as their contemporary variants, such as neo-colonialism. Edward Said (1993: 8) provides a useful distinction between colonialism and imperialism. Imperialism he defines as the practice, the theory and the attitudes of a dominating metropolitan centre ruling a distant territory. Colonialism, by his definition, is a specific articulation of imperialism associated with territorial invasions and settlements. Williams and Chrisman (1993: 2) also propose that colonialism is a phase within a more persistent process of capitalist imperialism, spanning until the present. Like Said, they define colonialism as that phase of imperialism in which the expansion of the accumulative capacities of capitalism was realised through the conquest and possession of other people's land and labour in the service of the metropolitan core.[2] Colonialism, then, entails the establishment and maintenance of domination over a separate group of people, who are viewed as subordinate, and their territories, which are presumed to be available for exploitation. This is clearly expressed in nineteenth-century British imperialism in which territorial expansion ensured that raw materials were supplied to the metropolis and new markets were created for manufactured goods (King 1990a: 49).

There is, then, an implied chronological ordering in the terms colonialism and imperialism. Said (1993: 8) proposes that 'direct colonialism has largely ended' but 'imperialism ... lingers where it has always been, in a kind of general cultural sphere as well as in specific political, ideological, economic, and social practices'. In part Said is referring to the tenacious persistence of the ideologies, practices and economies of high colonialism in the present moment of formal decolonisation. He is also alluding to various cultural and economic expansions which have continued well beyond the territorial appropriation associated with the colonial period. Neo-Marxist accounts of recent global capital accumulation, most notably Wallerstein's (1974) world-system theory, propose a global imperialism far more vital and comprehensive than Said's 'lingering' colonialism. Such theory outlines an international framework of accumulation based on corporate monopoly capital and an international division of labour. This world economy produces an order of uneven development based on the core, and the variously dependent semi-peripheries and peripheries.

The economic dimension of imperial expansions (be they colonial or 'newly' global) is undeniable, as are the uneven divisions of power and privilege they produce. But explanations that rest too heavily upon this logic work only to re-centre the metropole by incorporating 'the world', and all that happens in it, into the accumulative logic of the core. The contingency of this logic, its need to negotiate an often reluctant and almost always transformative periphery, is clearly lost within narrow political economy accounts (Shohat and Stam 1994: 17). World-system theory is an example of Marxism's capacity to generate a 'universal narrative of the unfolding of a rational system of world history' (R. Young 1990: 2). Such neo-Marxist accounts elucidate a 'world-system' in order to move towards the radical dissolution of that system, as in Jim Blaut's (1987, 1992a, 1993) anti-imperialist writings. But before the revolution – in a sense *for* the revolution – world-system accounts elaborate a comprehensive structure of Eurocentric diffusion to which alterity is dependently pinned and culture simply a vulnerable adjunct. Explanations which attend to the local cultures of imperialism open out the heterogeneous and mutable nature of these enterprises. Imperialism was activated by numerous desires and needs, colonialism took hold in a variety of forms and colonialist formations survive and are reactivated in a multitude of ways. Furthermore, at any one time imperialist visions are open to contest from within and without. Imperialism always produces an indeterminate array of formations, some of which are outside of its reach and work to unsettle its power.

Let me turn to the relevant example of the building of the British empire to elaborate the point about the internal variability of imperialism. Scientific and legal theories of social evolution gave British expansion across the world a 'natural' logic. The world, in evolutionary terms, was inhabited by 'advanced' and 'backward' peoples. As Said (1978: 207) notes, John Westlake's *Chapters on the Principles of International Law* (1894) advised that the 'uncivilized' sections of the globe should be annexed and occupied by the 'civilized' and advanced powers. Such racialised social constructs provided the 'sense' for social relations established under colonialism. That is, social evolutionary logic did more than just categorise the world's people in hierarchical ways, it also legitimated the exercise of power through these differences. Such categorisations were however part of a broader, western humanism which placed European Man as the centre of history and assigned to him the responsibility for the welfare of 'humanity'. As Robert Young (1990: 122) argues, drawing on the incisive work of Frantz Fanon, humanism was part of the legitimating drive of imperialism. But this was a conflictual concept predicated on the exclusion and paradoxical 'dehumanisation' of various feminised and racialised others. As Homi Bhabha (1990a: 71) notes, if the west followed its colonial project to the 'peripheries' it would there face its ambivalent doubleness as both 'civilizing mission' and 'a violent subjugating force'.

But it is not simply the internal paradoxes of imperialism that need to be registered. The types of social categorisations and adjudications I have just sketched did not realise themselves uniformly. They took an unruly passage from the heart to the edge of empire producing distinct forms of colonialism (Thomas 1994: 39). A permanent territorial colonisation might entail, at least in principle, the wholesale destruction of existing societies and the building of new societies modelled on the imperial core – as was the intention in Australia. Or it may have been designed to keep the indigenous society 'intact' but reoriented, in the service of the core – as was the intention in India (Taylor 1993: 18). As Thomas notes, the colonial conquest of India was founded on an appropriative incorporation of local knowledges and social structures which were considered distinct but still 'civilised' and useful. The colonisation of Australia, in contrast, was premised upon the view that indigenous land relationships were incommensurate with British notions of property. They were placed into the sphere of 'belief' rather than legally recognisable right, opening the way for the expropriation of this 'unowned' land (Blaut 1993: 25).

But even within one imperial system of categorisation there is often an internal dynamic of competing ideologies and moralities. For example, the categorisation of the Australian continent as a land unoccupied, *terra nullius*, was by no means a uniformly held view. Within the colonial territories and in the imperial core there was fierce debate about the rights of indigenous peoples in the Australian colonies. An 1837 British Commons Select Committee recognised that 'native inhabitants of any land have incontrovertible right to their own soil: a plain and sacred right' and noted, disapprovingly, that in the case of the Australian colonies this right was not recognised. *Terra nullius* is better understood as the fantasy of an emergent nation, part of a future-oriented reconstruction by colonists, rather than as the mark of the common sense of British imperial ideology. Even within the Australian colonies themselves there was an uneven acknowledgement of Aboriginal territorial rights: so while a treaty which presumed Aboriginal prior occupation was signed in the Port Phillip colony the validity of such 'deals' was rejected by the colonisers of New South Wales.

Thomas also notes that within any one 'mode' of colonialism there were important variations based on the colonising figure: be they explorer, missionary, trader, government official or a woman in their partnership or service. As he elaborates:

> the dynamics of colonialism cannot be understood if it is assumed that some unitary representation is extended from the metropole and cast across passive spaces, unmediated by perceptions and encounters. Colonial projects are construed, misconstrued, adapted and enacted

by actors whose subjectivities are fractured – half here, half there, sometimes disloyal, sometimes almost 'on the side' of the people they patronize and dominate, and against the interests of some metropolitan office.

(Thomas 1994: 60)

Imperialism, then, may have been energised by apparently quite rigid ideologies which centred the west and legitimated certain social relations of domination, but these ideologies tumbled into the fractured and erratic everyday practices of the personalities (colonist and colonised) who were forced together in the making of colonies. The overarching outcome of 'imperial domination' came to be through a broad range of encounters including stark oppression but also 'sympathy and congruence' as well as 'antagonism, resentment or resistance' (Said 1993: 47).

IMPERIALISM AND SPACE

Already this general discussion of imperialism and colonialism has unavoidably implicated considerations of space. In Edward Said's afterword to the 1995 reprint of *Orientalism*, he proposes that the task for the critical scholar of imperialism is to connect the 'struggles of history and social meaning' with the 'overpowering materiality' of the 'struggle for control over territory' (1995: 331–332). Imperialism for Said (1993: 271) is 'an act of geographical violence through which virtually every space in the world is explored, charted, and finally brought under control'. Imperial expansions established specific spatial arrangements in which the imaginative geographies of desire hardened into material spatialities of political connection, economic dependency, architectural imposition and landscape transformation.

The role of the spatial imaginary in the imperial project is perhaps most clearly evident in the spatial practices of mapping and naming. Harley has demonstrated the convincing relationship between cartographic practices and the production of 'known' space in imperial projects. The quest to map may well be undertaken as an 'innocent' cartographic science, but the maps produced never simply replicate the environment. They are part of the 'territorial imperatives of a particular political system', most notably that of imperialism (1988: 278). They are, as Harley (ibid.: 283) notes, 'the currency of political "bargains" ... the force of the law in the landscape'. As Huggan (1991) suggests, the cartographic exercise within the colonisation process depended upon a technique (and a hope) of representing a stable and knowable reality in what were unknown lands inhabited by unknown people.

The map has in a sense become the over-determined signifier of the spatiality of the imperial imagination. But the intersection of imaginary and material spatialities was present in other areas of the colonial project and no more so than in the making of cities. The successful exploitation of colonial resources required cities to be built in the colonies. These functioned as centres for colonial administration, sites of local production and consumption, and conduits for the flow of goods and services. There has been a long tradition of studying the colonial and the Third World city in the historical formation of imperialism (see, for overview, Simon 1984; Yeoh 1991; Alsayyad 1992; King 1992). One theme within this broad and theoretically diverse field is the transfer of European architectural styles and planning practices as part of the project of colonial domination (for example, Ross and Telkamp 1985; King 1990a). Other work, drawing on Wallerstein's world-system theory, has been concerned with describing the role of the colonial/Third World city in the emergence of a broader capitalist system of dependency (Friedmann 1966; and more recently Timberlake 1985). Anthony King (1976, 1990a), for example, has proposed a theory of colonial urban development to explain the political, economic and cultural processes that gave rise to new cities in colonised territories. Colonial cities were, according to King, important sites in the transfer of modern capitalist culture to new worlds. This can be seen in the architectural form and planning of such cities which regularly mimicked the cities of the imperial home. Colonial cities also operated as important sites in the deployment of the technologies of power through which indigenous populations were categorised and controlled. Here town planning became the mechanism by which colonial adjudications of cleanliness, civility and modernity were realised quite literally on the ground. Not least, it was in the name of the ideal city that many of the most comprehensive colonial territorialisations and displacements occurred and the most rigid policies of segregation were implemented (King 1990a: 9).

Colonialism did not simply involve the transfer of metropolitan processes of urbanisation to the colonies; there was reverse movement as well. As King (ibid.: 7) argues, 'urbanism and urbanisation in the metropole cannot be understood separately from development in the colonial periphery'. This involves more than the processes that brought exoticised fads to the architecture of imperial cities or saw monuments made to the triumphs of empire. The use of peripheral territories for primary production and resource extraction facilitated, indeed necessitated, the growth of industrialised and commercialised urban centres in the imperial core. The empire produced new social and material arrangements in imperial cities (Rex 1981: 3). Colonial mentalities and practices of categorisation and governmentality often folded into the management of the new underclasses of urban industrialisation. The magnitude of the transformations which occurred in imperial cities in turn produced their own divergent counter-tendencies, such as anti-urban movements. The pre-industrial nostalgias of, say, William Morris's socialism and the early movements to preserve the historic built fabric of the city, were also pre-imperial nostalgias and were often stridently expressed as such.[3]

Although King is sensitised to the culture of urbanisation and to its links with imperialism, his theory is situated within a world-system account of the history of globalising capital.[4] Urban space is particularly seductive in terms of the construction of world theories of capitalism. De Certeau (1984: 95) suggests that 'the language of power is itself "urbanizing"' and that the city stands as 'a totalizing and almost mythical landmark for socioeconomic and political strategies'. As Ed Soja proposes, it is as if political economy understandings of advanced capitalism lead inevitably to the city. Urbanisation becomes 'a revealing social hieroglyphic through which to unravel the dynamics of post-war capitalist development [in] an increasingly urbanized world economy' (Soja 1989: 94). Just as Soja constructs Los Angeles as the paradigmatic postmodern city, in a not-so-unrelated project Marxism constructs the city (First and Third) as the paradigmatic site of globalising capital.

World theories of urbanisation — be they cultural diffusionist or neo-Marxist — tend to deactivate space by seeing the city as the uncontested imposition of imperial territorial arrangements. It is not that such a thesis is wrong-headed, but more that the scope and the scale of this notion of imperial territorialisation moves away from the technologies of space which worked to embed imperialism. Brenda Yeoh (1991), using the example of Singapore, has shown that colonial cities are not simply produced by the smooth flow of imperial spatialities to colonised lands. Rather, she has shown how colonial aspirations of territorialisation depended upon fine-grained spatial technologies of power such as town-planning regulation and policing. Furthermore, it was this locally articulated spatial power that formed the focus for subversive and resistant activities by the colonised dwellers of such cities.

Paul Carter's approach to the spatial history of colonial settlement in Australia also gives some important clues as to how an activated spatiality may be incorporated into readings of colonial and imperial processes. For Carter (1987: 46) space creates history by inventing 'the spatial and conceptual co-ordinates within which history ... occur[s]': providing the agreed points of reference, the maps which define the architecture of 'here' and 'there'. Such ordering was especially evident in the planning of colonial cities around the spatial template of the grid. These plans placed a rational spatiality of urban order over the unknown ('irrational') spatiality of Aborigines/Nature. The transportation of the urban grid to the land of Australia signified the arrival of imperial authority but Carter rightly argues that the purchase of this authority was contingent and ambiguous. He does not wholly subvert the link between power and mapping/naming, but he does point to the uncertain stability of the hopes contained within the imperial spatial imaginary. If the mapping and charting of country was part of the material possession of people and places, then it became so because its fantastic scope was anxiously articulated on the ground, through settling and journeying over a page which was far from blank. These more unruly encounters worked to realise the perfection of imaginative projections; in Carter's language, to transform the 'haze' of (pre-modern) space into the 'clear outlines' of (modern/imperial) place.

For Carter (1987: xvi and xxii), history that speaks only of time on the deactivated 'stage' of space (space as an 'empty interval, a natural given') is imperial history. To activate space, to produce a spatial history, is fundamental to Carter's project of taking history beyond imperialism. Like many geographers, I see Carter's reworking of history as a landmark event or, to use Carter's preferred language, 'spatial occasion' (ibid.: 143). But reaching a truly postimperial or postcolonial perspective requires more than (re)activating the spatial narratives and imaginings within *past* projects of making empires. The challenge, it would seem, is to register this spatial sensibility in the present and to recognise that while colonialism attempted to carve 'clear outlines' onto the 'haze' of space, this has been an incomplete project. The diasporic movements, the insurgent claims for rights over land, the pervasiveness of imperial nostalgias, all point to the spatial 'haze' of the present.

THE LIMITS OF THE POSTCOLONIAL

Just as there is confusion around the terms colonialism and imperialism, so there is a proliferation of uses and implied meanings pinned to the term postcolonialism. The term refers not only to formal political status, but also to certain subject positions, political processes, cultural articulations and critical perspectives. Paradoxically, the process of decolonisation after the Second World War, which released former colonies from nineteenth-century colonial arrangements, is perhaps the least meaningful signifier of what might be thought of as postcoloniality. The move to formal independence is shot through with imperialism itself. Formal postcolonial status is a product of imperial cores conceding power over colonised territories. More often than not structures of neo-colonialism provided the very preconditions for such gestures of decolonisation (R. Young 1990: 122). Contemporary resettlements and reterritorialisations undo the geographies of colonialism. Yet diasporic groups, citizens of newly independent nations and indigenous peoples still face the force of neo-colonial formations and live lives shaped by the ideologies of domination and the practices of prejudice established by imperialism. Historically speaking, postcolonialism implies a liberation frequently beyond the limits of existing power relations.

The equivocal nature of postcolonialism as a formal political and historical condition is well demonstrated by nation-states like Australia. Australia might best be described as a 'break-away settler colony' which was founded on the principle of transferring imperial power, with little change, from the core to the colony itself (McClintock 1992: 89). Australia is not part of the Third World; it is a reasonably successful western nation which is, relatively speaking, economically independent of dominant core countries. Australia's historical and economic development, and its place in global capitalist relations as an advanced or First World nation, means that it is as much metropolitan as it is (post)colonial (Williams and Chrisman 1993: 4). Yet while Australia has always had tendencies which have moved it away from its colonial maker, it is still a member of the Commonwealth and the British Queen is still officially the head of state. Indeed as Australia struggles with the idea of becoming a republic, of finally realising a more complete and formal state of independence, it is evermore apparent that it is a nation deeply marked by forms of internal colonialism. This is clearly evident in the fact that Australia has a Fourth World, Aboriginal Australia, within its First World boundaries.

For the indigenous peoples of Australia, the 'post' of postcolonialism is still a long way off. Indeed in self-consciously multicultural nations like Australia, the original political nexus of coloniser and colonised relations is being decentred by the presence of a range of migrant settlers from around the world: Asia, Africa, South America, the Middle East and elsewhere. These new settlers are not without their own experiences of colonialism, but they are as often as not removed from the specificities of the British colonisation of Australia. 'Post'colonial politics in Australia is practised within a nation-state which joyously moves towards multiculturalism: where the often irreconcilable responsibilities produced by the colonialist violence of dispossession compete with the seductive promise of a more worldly and most appetising (for food is one of its great markers) multicultural Australia. It is in the face of the displacing force of multiculturalism that the claim of being indigenous, *not a settler of any sort*, has gained in importance in First Nation/Fourth World political movements in nation-states like Australia and Canada. The claim of habitation, of not having arrived, becomes the means by which indigenous struggles gain distinction in an increasingly diasporic present (Clifford 1994: 308). Australia, then, is the sort of nation that may be visioned as postcolonial by some but feels decidedly colonial to others. It is the type of ex-colonial territory that points to the formal limits of the historical condition called postcolonialism and the fantastic optimism of the 'post' in postcolonialism.

The example of Australia suggests that the historical moment in which imperialism or neo-colonialism are superseded is elusive. But this does not preclude the existence of a range of formations which may be postcolonial in intent or effect. In Australia postcolonial or counter-colonial formations exist within a nation-state which has not 'needed' to experience the historical gesture of 'independence' and where, indeed, 'independence' is simply not an option for the colonised. It is the existence of such conditions which led Mishra and Hodge (1991: 407) to make the useful distinction between what might be thought of as overt 'oppositional postcolonialism', found in post-independent colonies and nationalist claims, and a 'complicit postcolonialism' which is a type of 'symbiotic postcolonial formation' which exists as an ever present 'underside' in settler colonies like Australia.

If the term postcoloniality is refracted onto the imperial heartlands it is further problematised. London may well be the city of the metropolitan core of a now dismantled British empire, but there is much to suggest that the foundational ideologies of imperialism live on in this city; shaping contemporary economic status, local class divisions and racial politics, and nationalist articulations. Here the traces of imperial might continue to condition the ways in which cities like London reorient themselves within new global and regional arrangements. London, for example, can claim its status as a global city in a new global and regional economic order precisely because it is able to elaborate functions developed during its nineteenth-century colonisations of the world. In cities like London the unprecedented levels of postcolonial migrations and settlements have produced diasporic communities drawn into familiar, but now localised, arrangements of power. Indeed, many of the new labour arrangements of global cities, including London, quite literally rework people already categorised as available for exploitation under colonial economies. Such diasporic communities have varying investments in the British nation-home. Their sense of identity is constituted from multiple localities: Bangladesh as much as Britain, Sylhet as much as Spitalfields, to gesture to the relevant case developed in Chapter 4. In this new 'ethnoscape' (Appadurai 1990: 7), ambivalent new communities are thrust together with anxiously nostalgic old ones. Xenophobic fears and gentler fantasies of a surer past of imperial might manifest as a politics of racism, domination and displacement which is enacted, not on distant shores, but within the very borders of the nation-home.

Thus far, I have pointed to the limits of the formal condition of postcolonialism which suggests that the term is often 'prematurely celebratory'

(McClintock 1992: 87). During (1992: 340) suggests that 'fröhliche (joyful) postcolonialism' denies the ongoing efficacy of imperial structures of power – the way in which the imperial desires (past and present) are in the here and now. Indeed, it might be useful to think, as Bhabha (1994: 6) does, that the existence of a postcolonial politics is not a mark of being beyond colonialism but precisely a 'reminder of the persistent "neo-colonial" relations within the "new" world order'. If this is the case, then where is the postcolonial?

Postcolonialism may be better conceptualised as an historically dispersed set of formations which negotiate the ideological, social and material structures of power established under colonialism. Ashcroft *et al.* (1995: 7) go so far as to say that there can be no postcolonialism without the historical precondition called colonialism. They are, in a sense, stating the obvious: one is, of course, the determining condition of the other – one is always already 'contaminated' by the other (Hutcheon 1995: 134). From the beginnings of colonial encounters to the contemporary moment of a disseminated global imperialism, there have been counter-colonial movements and outcomes which unsettle colonialism. These may be registered as anti-colonial nationalisms but also as more fractured forms of opposition and destabilisation.

Frankenberg and Mani (1993: 295) suggest that the term postcolonialism necessarily and problematically has globalising tendencies.[5] They question the way in which the term is so closely linked to the writings of 'first-generation diasporic intellectuals', so that the experiences of these writers have come to stand for the nature of postcoloniality across a diverse range of settings. The real trouble with the term postcolonialism is perhaps better diagnosed by McClintock (1992: 86) who argues that it generalises diverse historics and links them once again, even if in counterflow, to the European core. This is to hint at the possibility that those articulations and practices that are subsumed within the notion of the postcolonial may not be about the colonial at all – may even be blithely indifferent to it. In this sense the marking of the practices and articulations of the Other as postcolonial is to provide them with an implied strategic sensibility, an agency, which is always produced by, for or against the core. McClintock sees this as a process whereby the various cultures of colonised subjects are hauled back into knowability by way of their refracted (negative) relationships to the centre. McClintock, then, is implying that there can be an alternative way of knowing Otherness, a way of unleashing 'the world's multitudinous cultures' into an independent space of 'positive' distinction. Seductive as this notion might be, it is in part driven by a certain nostalgia. It is hard to imagine a space that is not somehow touched by colonialism or, for that matter, to imagine cultures that are not somehow constituted out of their necessary positioning in the modern. If we are not yet past the last 'post' we are most certainly past the last of the lost tribes. As

Appiah (1992) notes in the context of African cultural production, 'we are already contaminated by each other ... there is no longer a purely autochthonous pure-African culture awaiting salvage ... just as there is ... no American culture without African roots'. To imagine this might be so is to reinvigorate a modernist binarism which, Appiah says, we must learn to live without.

These are, of course, all issues to do with the nature of postcolonial theory, as much as they are to do with the nature of a postcolonial condition. Postcolonialism as a theoretical and analytical perspective includes a diverse range of perspectives: historically based critiques of colonial discourses, anthropology's critical revision of its own colonial complicity, accounts of formations such as diasporas, studies of the cultural productions of colonised peoples and, not least, the various articulations of those who are themselves speaking from the margins. Disparate though postcolonial critical studies are, they share a common political project (albeit unevenly realised) which is counter-colonial. Anne McClintock argues that the intent of postcolonial theory is to challenge the logic of linear 'development' and its 'entourage of binaries'. But she also points out that the term postcolonial paradoxically re-establishes a binary orientation, a return of the colonial. The term postcolonial is 'haunted' by the logic of western historicism and while it 'heralds the end of a world era' it does so 'within the same trope of linear progress that animated that era' (McClintock 1992: 85).

Such a tension is found in the work of Homi Bhabha, whose postcolonial project includes a radical deconstruction of the Self/Other binary. Bhabha's theorisations are against colonialism but his project has not consistently been one of activating Otherness in an overtly counter-colonial mode or tracing the formations of intentional resistance. Bhabha is concerned with colonialism's own vulnerability to itself. Through a psychoanalytic approach, Bhabha elaborates the fantasised subject, the latent or unconscious Other, of colonialist constructs. He proposes that the rules of recognition contained within the notions of Self and Other, upon which colonial power was built, do not produce a clear command of authority over difference or even a successful repression of difference. Rather, these constructs always produce an excess, something which slips outside of the binary order. This uncontainable surplus establishes the very basis for the disallowance of colonialism's own authority, and this ambivalence establishes the necessity of colonialism's anxious repetition of the stereotypes which 'allow' (but also always defer) mastery. Bhabha elaborated this idea first through mimetic modes in colonised subjectivity, that is, the Other who mimes the Master. Rather than mimesis providing proof of the realisation of the civilising intent of colonisation, it establishes a partial and distorted representation which menaces the coloniser more than it comforts. In this sense, Bhabha undoes the assumption of Said's (1978: 77) *Orientalism*; that colonial discourse is 'possessed entirely by the coloniser'. Bhabha replaces calm mastery with an agonistic space of colonial authority.

There is little doubt that Bhabha's psychoanalytic perspective on colonialist discourses has productively elaborated the critical tradition inaugurated by Edward Said. In so far as Bhabha's understanding of the ambivalence of colonialism dislodges the surety of colonial power, his analytic perspective is postcolonialist. Yet Bhabha's postcolonial analysis has not gone uncriticised. Thomas (1994: 49 and 51) suggests that attention to the 'psychic dynamics of self–other relations . . . cannot be accorded . . . historic peculiarity' and calls for a more 'socially and historically grounded characterisation' of colonialism's cultural purchase. Thomas implies that the transcendental applicability of the psychoanalytic perspective will be challenged by historical specificity. Robert Young (1990: 146) also argues that Bhabha paradoxically deploys an analytic frame without reference to its Eurocentric historical provenance. These critics are suggesting, then, that Bhabha reproduces a colonialist gesture in the psychoanalytic infrastructure of his critique of colonialism.

The postcoloniality of Bhabha's work has also been called into question by its own ambivalence towards the agency of the colonised subject. The presence of agency or intentionality is, for some, a crucial component in designating certain formations as properly postcolonial (Slemon 1991). For example, in his discussion of African cultural production Appiah (1992) proposes that postcoloniality means more than the traceable intermingling of African and colonial cultures, a version of hybridity. For him the 'post' in postcolonialism refers to the 'space-clearing gesture', that is, those formations which are actively concerned with going beyond colonialism. Within Bhabha's reading of colonialism, *post*colonial effects are produced by the agonistic ambivalence of colonialism itself rather than the agency of the colonised. Robert Young (1990: 147) argues that even Bhabha's notion of mimicry – in which the colonised subjects take up the guise of the coloniser – places agency within the equivocal circulation of *colonial* constructs. In Bhabha's understanding, the mimetic performance of the colonised subject subverts colonialism not because it might be a conscious act of (mis)appropriation, but because it has a menacing effect which is produced by colonialism's own paranoia. This is, then, as Young puts it, 'agency without a subject'.

While Bhabha's notion of mimicry proposes a colonial absorption of agency, his concept of hybridity attempts to return it to the colonised. In his essay 'Signs taken for wonders' (1985), Bhabha's concern turns to the way in which colonial ambivalence produces hybridisation, an inevitable 'splitting' of subjection. Hybridity is not just a mixing together, it is a dialogic dynamic in which certain elements of dominant cultures are appropriated by the colonised and rearticulated in subversive ways. In Bhabha's words, hybridity is about the 'seizure of the sign ... a contestation of the given symbols of authority' (1992: 63). Such subversions are uncanny returns, where disavowed and repressed subjectivities and knowledges 'enter upon the dominant discourse and estrange the basis of its authority – its rules of recognition' (Bhabha 1985: 156). The concept of hybridity implies that postcolonial effects are no longer only unconscious by-products of colonialist constructs. They are the creative remaking of the colonial past by the colonised in the service of a postcolonial present/future. Through hybridity a postcolonial effectiveness is returned to the colonised, who steer a subversive return to the colonial heart.

Both Robert Young (1990) and Benita Parry (1987) have noted that Bhabha's recovery of the agency of the colonised is still often found 'between the lines' of colonial discourses. Their responses allude to the same trouble: that the various documented histories of overt resistance to colonialism are displaced by articulations of subversive excess which too closely inhabit the colonial. Put bluntly, and perhaps too simply, this is because Bhabha's main concern is with the field of colonial discourses rather than anticolonial discourses and formations. As Williams and Chrisman (1993: 16) point out, all perspectives of this type position colonial subjectivity as having 'primacy' and native or subaltern subjects 'as secondary "subject effects" ... within the discourse of empire'. These critiques are thus concerned with the failure of Bhabha's analysis to do justice to the politics of contestation, expressed, at the very least, as tactics of subversion but at their most striking as collective insurgencies. To haul the dubious politics of Bhabha's colonial excess back into contact with the politics of agency is a necessary (moral) manoeuvre in the pursuit of an anticolonial perspective.

The challenge is surely not to deny that colonialism produces its own vulnerability – or to say historical specificity will displace (or disprove) Bhabha's perspective. Rather, it is to see that Bhabha's notion of the unsettling (postcolonial) surplus of colonialism is a component in a wide range of postcolonial formations which are expressed in variable ways and in distinct settings. Bhabha is right to point to the instabilities of colonialism produced by the hazy subversions made possible by the ambivalence of colonial discourse itself. But these instabilities can also arise in sharper counter-colonial movements of, say, certain nationalisms which may well mobilise essentialist notions of a precolonial identity. Furthermore, all negotiations of identity are located within very specific hierarchies of power and particular political and economic frames. As Chandra Mohanty argues, in the context of the struggles of Third World women, identity-based understandings of domination need always to be located in the material politics of everyday life (Mohanty 1991a: 10–11). It is not solely that discursively constituted notions of identity have material effects, but that the sheer uneven materiality of the lives of people affected by imperialism must inform the moral and ethical function of critical postcolonial studies.

POSTMODERN SPACE AND THE (POST)COLONIAL

Williams and Chrisman (1993: 13–14) note that the much-debated relationship between postcolonialism and postmodernism still requires 'lengthy and careful delineation'. This nexus is of particular relevance to this volume which takes as its empirical focus not the colonial past but an urban present which is so often read as a paradigmatic site of the 'condition of postmodernity' (for example, Harvey 1989; Soja 1989). There is little doubt that postcolonial theoretical revisions have productively cross-fertilised with postmodern theory. The postmodern projects of deconstructing Master narratives, unsettling binaries and admitting marginalised knowledges, follow closely the objectives of the postcolonial critical project. Similarly, these various perspectives are conjoined in their attention to the relationship between discourse and power, the socially constituted and fragmented subject and the unruly politics of signification – the workings of irony, parody, mimicry (Ashcroft *et al.* 1995: 117). As theories, postmodernism and postcolonialism appear to be one and the same, with postcolonialism perhaps distinguishing itself by the primary concern it has with the processes associated with the condition called colonialism, as feminism has with patriarchy. This overlap between postmodern and postcolonial theoretical perspectives has produced its own discourse of differentiation in which claims are made as to which came first, which has the 'real' politics, and what one does to the other.

Linda Hutcheon, for example, has argued that postcolonial politics, like feminism, cannot enjoy the luxury of poststructuralist engagements with decentred subjectivities (Hutcheon 1995: 130–131). She points to the necessity of, at the very least, momentarily fixed ideas of identity and difference in political projects which seek to challenge hegemony and affirm 'denied or alienated' subjectivities. Crapanzano (1992: 90) mounts an even stronger argument against postmodern theoretical perspectives by proposing that the painful history of silencing, in other words colonialism, precludes a postcolonial politics of 'playfulness' based around surface, spectacle, polyphony and heteroglossia. Crapanzano here seems to underestimate the political potential of the phenomena he categorises as 'play'. As Hutcheon herself notes, postcolonialism does articulate itself through the 'play' of irony, mimicry and hybridity and these strategies/effects are powerful in their capacity to unsettle the colonial (see also R. Young 1995).

Others have argued that counter-colonial and non-metropolitan cultural productions have been appropriated into postmodernist deconstructions of western hegemony (for example, Adam and Tiffin 1991: viii). This suggests that a postmodern sensitivity to difference is simply another form of Eurocentric expansionism in which the experiences and cultural productions of the marginalised now become a means by which the core understands/deconstructs itself. Gayatri Spivak (1988b) of course elaborates a version of this contradiction in her important essay 'Can the subaltern speak?'. Here she outlines the political difficulties she has with poststructuralist attempts to deconstruct Master Narratives by gesturing towards an often frighteningly undifferentiated Otherness. This, she argues, establishes the 'assimilationist' conditions within which Otherness can be spoken and marks an 'epistemic' articulation of the violence of the west (ibid.: 280). Here Alterity is elaborated en route to reproducing the 'West as Subject', albeit by registering the inappropriateness of its own past.

Spivak's complaint about poststructuralist theory is not simply that it makes use of Otherness in its own interests. She also complains that poststructuralism's aversion to grand narratives means that it is incapable of dealing with something as bold as global capitalism, something she, like Mohanty, still considers vital to understanding the nature of power and privilege in the (post)modern world (ibid.: 272). Interestingly enough, if Spivak were to seek comfort from some of the most active global theorists of the postmodern condition she might find, as feminists Rosalyn Deutsche (1991) and Meaghan Morris (1991) did, a most unfamiliar scene.

Recent accounts of contemporary global capitalism have proposed that the old arrangements of core and periphery have been radically rearranged in more dispersed, more disorganised, more flexible, but unfailingly comprehensive, forms of capital accumulation. According to Jameson (1984, 1991: 157), monopoly capitalism (imperialism) has 'mutated'. Within this new regime of capital accumulation, imperial invasions (presented most plainly as technology-assisted commodifications) are fast, all-pervasive and decidedly multinational. For Jameson, then, postmodernity is the moment when 'old imperial maps have been lost' (R. Young 1990: 117) and we are cast into the disorientations of an unmappable landscape of 'great global multinational[s] and decentred communicational network[s]' (Jameson 1984: 84). Indeed, Jameson's own anxieties about this new spatial and social complexity, the unmappability of the present, lead him to propose a 'cognitive' remapping of the 'world space of multinational capital' (Jameson 1984: 89).

David Harvey's (1989) 'mapping' of post-fordist capitalism presents it as marked by global economic restructuring, increasingly disintegrated divisions of deregulated labour and proliferating aestheticised economies of consumption. At one level Harvey argues that conventional spatialities of capital have experienced a radical transformation under the time–space compression of postmodernity; producing new regional trade and political alignments, new flows of information and new patterns of mobility and settlement. The plodding colonial possession of territories, it seems, has been replaced by the marauding logic of Late Capitalism. But this logic still has remarkably familiar spatial and social features. Multinational industrialisation has replaced monopoly industrialisation but it elaborates a familiar geography of exploiting cheap Third World labour and resources. Within former core countries, manufacturing is replaced by service and consumption-based industries which often, quite literally, take up the spaces abandoned by factories and the working class. Harvey's version of postmodernity is as familiar as it is strange.

Within such narratives of global restructuring the First World city holds a special place, both as the site and signifier of such change. Indeed, postmodernity can lay claim to a new type of city, the 'global' city (Friedmann and Wolff 1982; Sassen 1991). Global cities – such as Los Angeles, London, Tokyo and New York – are made by their control of expanded, complex transnational financial services. They are the nodal sites for global banking and financial operations and the homes for the headquarters of multinational corporations. Of course, postmodernity is not confined to such super cities and all kinds of First (and Third) World cities are now read as displaying the features of Late Capitalism's accumulative ways; such as spectacular sites of consumption, architectural pastiche, gentrified neighbourhoods and manufacturing sites reinvented as tourist destinations.

Keith and Cross (1993: 8) argue that the exoticised postmodern city will often celebrate ethnicity but processes of racialisation and racism remain 'a taboo vestige of colonial and neo-colonial exploitation'. Yet, as they go on to argue, even the most postmodern of cities has an 'architecture of power' structured around racialised values. The cities of modern South Africa, for example, were produced by colonialism and pose a threat to its stability. The racially diverse groups brought together in these cities have been subject to rigorous and ongoing spatial and social regulation through apartheid (D. Smith 1991; Crush 1992). In British cities, too, the relations of domination and exploitation established under colonialism determined the nature of postwar migrations and settlements: that is, who came to settle in the contemporary city; how they were incorporated into local labour markets; and the conditions in which they live (Rex 1973; Rex and Tomlinson 1979; Miles 1982). In contemporary British cities the elaboration of imperial constructs of difference has given rise to spatially segregated, racialised geographies of disadvantage (Gilroy 1987; S. J. Smith 1989, 1993). Such examples suggest that the racialised controls of colonial and imperial cities remain cogent features of contemporary cities, be they formally postcolonial or postimperial.

Keith and Cross (1993) suggest that some of the most influential accounts of postmodern cities contain race as an 'unspoken' silence. But it is not that Harvey or Soja, to take but two of the more influential scribes of the postmodern city, forget race. These authors do tackle the racialised nature of both the aesthetic and labour regimes of the postmodern city. Indeed one of the social phenomena upon which the narrative of urban postmodernity is constructed is that of social polarisation – the emergence of a more broadly defined (more than class) and sharply differentiated distinction between those who have and those who do not. Lash and Urry (1994: 145–146), for example, argue that there has been the emergence of a 'new lower class' into which 'large numbers of immigrants flow'. Within social polarisation arguments the complex politics of race is translated into a variant form of class differentiation produced by the now more thoroughly globalised and deceptively aestheticised unevenness of capitalism. A fractured, positional and often angry politics of difference is (mis)recognised as a static, structural outcome of advantage and disadvantage. Through this manoeuvre, the politics of race is cast off from the history of the constitution of difference and racialised subjects are denied the kind of agency captured by theorisations of a politics of identity. It is not simply that there is not enough race in these accounts of the postmodern city, it is that the cultural politics of racialisation is deactivated.

In contrast, Paul Gilroy sees the conflictual space of the city as deeply entwined with a cultural politics of race:

> We must confront the extent to which the cultural politics of 'race' reveals conflict over the production of urban meanings and situate the meanings which have already been identified as constitutive of 'race' in their proper place as contending definitions of what city life is about.
>
> (1987: 228)

It is undeniable that the condition of postmodernity has meant that even the most recalcitrant Marxists have admitted culture, if not a cultural politics, into their understanding of the workings of capitalism. Culture is, as Jameson tells us, the 'new logic' of capitalism. Late Capitalism has produced, we are told, a uniquely 'semiotic' society which is 'regulated' by a combination of the material and the representational (Lash 1990). For Jameson there is hardly an area of life (not Nature, not the Unconscious, not Culture) untouched by the prodigious expansion of capital (Jameson 1991: 48–49). Within Jameson's narrative, a formerly autonomous cultural sphere is appropriated into the service of capital's accumulative logic and, once entrapped, delivered as an inauthentic culture of pastiche, simulacrum and commodification. Culture, at least in the First World of Jameson, serves not the processes by which identity is articulated or negotiated, but the (now) semiotic motor of capital accumulation.

Jameson depicts a deactivated cultural sphere in which difference is subsumed with the seamless homogeneity of 'hyperspace' (R. Young 1990: 204 n. 49). But his notion of homogenisation at the hands of a predatory capitalism is deeply problematic. In other accounts of postmodernity, late capitalism is depicted as being in a complex and contradictory arrangement with the local. It is undeniable that globalisation is occurring but, in a seemingly paradoxical reverberation, place specificity and social difference are being articulated as strongly as ever. Michael Watts (1991: 10) proposes that globalisation does not 'signal the erasure of difference' but rather the reconstitution and revalidation of 'place, locality and difference'. So, for example, recent economic restructurings regularly cohere around specific cultural differences in labour markets (see, as an example, Massey 1984). Similarly, different localities often elaborate distinctive self-images as place-selling strategies (see, as examples, Sorkin 1992; Kearns and Philo 1993 and Urry 1995). Such examples point to the way postmodernity manufactures difference in the service of its own consuming passions (see Urry 1995). This is not a productive politics of difference.

The inability of these hyperspaced versions of postmodernity to accommodate an activated cultural politics may well rest in their nostalgia for the uncontaminated Other and for its role in a revolutionary displacement of capitalism.[6] But as Arjun Appadurai (1990) argues, globalisation is not predatory in a simple sense. Products brought into the Third and Fourth Worlds are regularly indigenised, whereas indigenous products that spiral out of their local sites of production and into international markets as often as not perform a constructive role in counter-imperial articulations, allowing them to be amplified in the global context (ibid.: 15). Appadurai is proposing not only that globalisation does not obliterate the local but that it may also help to establish new globalised conditions for the expression of difference. His postcolonial account of postmodernity serves to illustrate that if an activated politics of cultural production and identity articulation were factored into accounts of the condition of postmodernity, slightly different stories would then begin to be told.

Jameson's (1991: 48) hope of 'positioning ... the cultural act outside of the massive Being of capital' takes him on a nostalgic journey to a world which is no longer. Chandra Mohanty (1991a: 2), in contrast, proposes a spatially diffuse racialised and gendered marginality which is no longer geographically confined (she is talking of the Third World) but exists in ambiguously connected diasporas and in complexly formed resettlements across the globe. Such spaces of Otherness are increasingly hard to incorporate into cohesive visualisations such as those proposed by Jameson. They are present within complex geographies of movement and resettlement where the place of the body may or may not be co-present with transnational spaces of ethnic, tribal, religious or even national identity. They are present within political terrains where localised identity might be positively amplified through the globalising forces of modernity. They are present within anticolonialist gestures which refuse the gaze of the west. Not only does Jameson aim to produce a new 'cognitive map' of postmodernity which is nostalgically imperial, it may be that the disordered geographies of the present are no longer available for the 'critical distance' which precedes the map. A cosier perspective, producing more detailed renderings of the local, is precisely what is needed to understand the ways in which the global and the local, the Firsts and the Thirds (as well as the Fourths), are mapped together into the 'leaky habitats' of the contemporary city (Chambers 1994b: 245).[7]

IDENTITY, THE PAST AND CITY SPACE

The nature of colonialism and postcolonialism compels us to think in ways that break the local/global binary and to speak of a micro-politics of place in which the two 'cohabit' (Massey 1993a: 64). The local is an active and constitutive force in the formation of social categories and the uneven operations of power between them. This is not simply a politics that is against globalisation. Nor is it simply a return to origins. Imperialism, in whatever form, is a global process – it occurs across regions and nations – but even in its most marauding forms it necessarily takes hold in and through the local. The embeddedness of imperialist ideologies and practices is not simply an issue of society or culture but also, fundamentally, of place. This can be seen in the way in which racialised constructions are, often quite literally, made through place. Kay Anderson, although not discussing imperialism *per se*, has argued that in First World cities the maintenance of hegemony over racialised categories works in and through socially constructed place (K. Anderson 1991: 28).[8]

Imperialism is also attenuated by the local. This is most starkly expressed in anticolonial formations which activate prior claims over the local (indigenous rights) or struggle to wrest back control of the local (as in some nationalisms and regionalisms). In such anticolonial struggles the local may well be evoked as some pure residual, untouched by the force of imperialism.

In such formations, the local is used to naturalise socially constructed identity and to reinscribe marginality in counter-hegemonic ways (Probyn 1990: 178). The articulation of identity in and through place is, then, not simply a given but also always a strategic, political fixing (Massey 1994: 8). That is, while nationalist or indigenous claims may draw on notions of origins and essence, this continuity and fixity is in part produced by the political context itself. And it is just as likely that such claims of origin will now be amplified through globalised networks of communication and affiliation. Claims of origin are one point in a contingent and negotiable terrain of positionings. In this sense, if there is a mappable politics of identity which is linked to place, then it is *also* always a rather less easily mapped 'politics of location' (hooks 1991: 145; see also Keith and Pile 1993a: 6).

It is precisely in the local that it is possible to see how the past, including imperial and pre-imperial pasts, inheres in place. This is not an archaic residue, but an active and influential occupation. A pertinent example of this is given by places that are designated as heritage, such as historic buildings or other cultural sites. These are inherited artefacts but they gain an active influence in the present by way of the various popular meanings and official sanctions ascribed to them. The making of heritage is a political process. Certain places may be incorporated into sanctioned views of the national heritage while others may be seen as a threat to the national imaginary and are suppressed or obliterated (see Hobsbawm and Ranger 1983; Wright 1985; Colls and Dodd 1986; Samuel 1989). Other oppositional places may be sanitised and depoliticised in their transit into officially sanctioned heritage. Which places do or do not become part of heritage and what transformations places undergo in this process of recognition is a key arena for combative struggles of identity and power. It is not simply that heritage places symbolise certain values and beliefs, but that the very transformation of these places into heritage is a process whereby identity is defined, debated and contested and where social orders are challenged or reproduced (Karp 1992: 5). Heritage is not in any simple sense the reproduction and imposition of dominant values. It is a dynamic process of creation in which a multiplicity of pasts jostle for the present purpose of being sanctified as heritage (Bommes and Wright 1982: 265; Wright 1985: 129).

In much recent analysis of postmodern urban landscapes, heritage has become the paradigmatic example of the way in which culture is the logic of Late Capitalism. The gentrification of historic inner-city neighbourhoods, architectural historicism and sites of heritage consumption (such as shopping malls and theme parks) are frequently cited examples of capital's voracious and clever appetite for the historicised local (Jager 1986; Zukin 1986; Hewison 1987; Thrift 1989; Urry 1990). Sharon Zukin, for example, argues that such processes of urban renewal and revalorisation dismantle 'older urban solidarities', grounded in locality-linked production, and replace them with consumption spaces 'shaded by new modes of cultural appropriation' (1992: 221). In this view, the local or vernacular is estranged from its authentic point of origin by the appropriative force of the global and becomes simply 'the spectacle of history made false' (Boyer 1992: 204).[9] Zukin (1991: 12), for example, goes so far as to argue that 'as markets have been globalised, place has been diminished'.[10]

The dichotomy of the authentically local and the appropriative global has its own problematic nostalgia. At best, the residual integrity of the local provides the hope of resistance. At worst, the local is seen as succumbing to the global, a compromised space of accommodation (Marcus 1992: 313). The focus on global appropriation of the local effaces the contradictory ways in which the past holds in the complex terrains of capital and power that constitute place. Harvey (1993: 11), in considering the 'authenticity' of meanings associated with dwelling and the mediated meanings associated with commodification, suggests that these are 'oppositions that contain the other'. Although this helpfully moves the understanding of identity and place away from a Heideggerean essentialism, it is hard to imagine which meanings or relationships might be unmediated (I. M. Young 1990: 233). Indeed, it may be that the distinction between the vernacular (authentic) and the commodified (appropriated) is no longer plausible. As Jackson (1991: 225) notes, it is precisely the paradoxes of 'appropriation' and 'authenticity' that hold the most potential for understanding the workings of heritage in the contemporary city.

Massey (1993a: 64) notes that the contemporary emphasis on sense of place (in which she includes the making of heritage) is problematic because it artificially localises place and draws around it boundaries that are not and never really were there. Dichotomies between original (presumably local and uncommodified) history and appropriated (presumably globalised and commodified) heritage may well need to be replaced by a more dynamic concept in which one always already inhabits the other. While heritage politics is more often than not about the local, about specific places, it is by no means a process which is sealed off from broader spatialities. Sanctioned heritage is taken up into national imaginings. Local sites are connected into global processes of commodification. Specific land rights struggles can influence pan-national and global indigenous rights movements. Neighbourhood territorialisations can form the basis of re-forging identity across the national borders of global diasporas. The politics of identity is undeniably also a politics of place. But this is not the proper place of bounded, pre-given essences, it is an unbound geography of difference and contest.

NOTES

1. Of course positional definitions of this kind are intensely fraught, as some of my own remarks in the Preface have already indicated. So, for example, in settler dominions like Australia, it is the colonist who is imperialist, but as an agent (often originally as a reject) of the imperial heart, the settler-coloniser claims a more parochial than global purchase on this power.
2. Shohat and Stam (1994: 15) make an interesting digression from these definitions. For them colonialism is the historical constant and imperialism the period of post-independence economic expansion.
3. For example, in his anti-urban novel *News from Nowhere* (1986 [1890]), William Morris makes specific reference to the world economy and its role in producing the city he writes against.
4. Anthony King's earlier work (1976) is quite sensitised to the contingency of colonial urban constructs. Similarly, much of his more recent work is clearly alert to the increasingly complex nexus of identity and place in contemporary cities shaped by imperial and colonial pasts. But a significant theme in much of his work has been to link the development of colonial, imperial and 'global' cities to world-system theories of urban development.
5. An example of the expansive definitions which adheres to the term postcolonial is Ashcroft, Griffiths and Tiffin (1989: 2), who propose that postcolonialism includes 'all the culture affected by the imperial process from the moment of colonisation to the present day'. Here everything the colonised subject does after the event of colonisation becomes part of the doubled space of being both only and always a product of colonialism and somehow automatically against colonialism.
6. This investment is clearly articulated in Jameson's explicit account of the relationship between 'Third World literature' and 'the era of multinational capitalism' (1986). This essay has been passionately criticised by Aijaz Ahmad (1992) for implying that the choice for Third World people is between nationalism on the one hand and globalisation (postmodernity) on the other. Not only does this obliterate the multitude of other possibilities available for Third World and other people who are attempting to respond to globalisation (not all postcolonialisms are nationalist), it also supposes that there is no possibility of positive engagements with globalising structures of communication and commodification.
7. King (1990b: 82) argues just the opposite of this. He says that to understand cities it is necessary to leave the 'cosy viewpoint' of 'within' and embark on a more 'uncomfortable' project of 'seeing them from outside'. From a perspective sensitive to the theoretical and political concerns of postcolonialism, the reactivation of the panoptic perspective is indeed 'uncomfortable'.
8. Anderson's work is one of the first sustained spatialised adaptations of Said's notion of Orientalism.
9. See also Keith and Pile (1993a: 7 and 9) for a discussion of Zukin's conceptualisation in the context of the relationship between identity and place.
10. This parallels Meyrowitz's (1985) notion that there is now 'no sense of place'.

3

NEGOTIATING THE HEART

PLACE AND IDENTITY IN THE POSTIMPERIAL CITY

•

logic has its limits and . . . the City lies outside of them.
(Royal Commission on Local Government in Greater London 1962)

In the early 1990s the British Law Lords reached a decision which allowed for the development of a relatively small parcel of land located on Bank Junction in the centre of the City of London. Without doubt this is prime real estate. Five major roads intersect at Bank Junction which, among other things, is the site of Mansion House (home of the Lord Mayor of London), the Bank of England and the Royal Exchange. The intersection has the look and the feel of a hub, and its grand buildings stand as monuments to the City's historic centrality to financial and commercial practices in Britain. In 1904 Neils M. Lund depicted the thriving bustle of this city space in his painting entitled *The Heart of the Empire* (Figure 1). Then, as now, the intersection was a symbolic site of a Britain made great by its global reach. Today it is an imperial space in a postimperial age. The decision by the Law Lords effectively ended a development struggle which had begun in the 1960s, which had seen some fifteen years spent in property acquisition, two schemes commissioned from leading architects (neither of whom have lived to witness the fortunes of their designs), and two gladiatorial public inquiries in which heritage redevelopment as opposed to new build development was the central issue.

The struggle over this symbolic heart of empire resulted in a prime piece of real estate being locked out of one of the most rapid and dramatic periods of restructuring and property speculation ever seen in the City and its surrounds. In the case of London, the Docklands/Canary Wharf redevelopments have rightly gained much critical attention as examples of mega-scale reinvestment in places previously under disinvestment (see, as examples, Brownill 1990; Wilson 1991; Coupland 1992; Bird 1993; Keith and Pile 1993a). At a different scale, attention has been given to the gentrification of inner-city spaces to meet the residential and lifestyle needs of new social and class formations associated with restructuring (see, as examples, Wright 1985; Samuel 1989; Wilson 1991). In a context of such palpable change it may seem contradictory to focus on a development site that was locked in controversy and where change was struggling to happen rather than occurring at breakneck speed. The planning saga which forms the focus of this chapter speaks of a space of desire, a place in a struggle between 'becoming' and 'remaining' (Pred 1986; also Pryke 1991 with specific reference to the City of London). The controversial redevelopment proposals for Bank Junction and the politics surrounding their 'absence' from the urban terrain of the 1980s City of London may say as much about recent urban transformation as those changes already realised.

In this prolonged redevelopment saga, Bank Junction (past, present and proposed) was invested with meaning by a range of interest groups: the developer, the local authority for the area (the Corporation of London), various local businesses and, not least, the powerful conservation lobby groups. The discourses generated by the planning controversy were not simply about the form and the function of this section of the City of London, a battle of old versus new or small business against big business. This highly publicised planning controversy became a nodal point in the imaginative reaffirmation of the identity and status of the City in relation to the nation and the rest of the world. In the span of half a century the City of London had gone from

FIGURE 1 Neils M. Lund's 1904 painting, *The Heart of the Empire*, depicts Bank Junction as the monumental, thronging hub of nineteenth-century imperial might. The painting takes an aerial perspective from above Hawksmoor's St Mary Woolnoth and looks westward past Mansion House (on the left) towards St Paul's Cathedral. The triangular group of nineteenth-century commercial buildings visually link St Paul's to Bank Junction by appearing as an extension of the Cathedral itself. These are the very buildings that were demolished during redevelopment. (Reproduced by courtesy of the Guildhall Art Gallery, City of London)

being the centre of an empire with a global reach, to one of the few urban centres given the privileged designation of 'global city' (King 1990b, 1991a; Sassen 1991). The City of London of the 1980s was both a postimperial city and a 'postmodern(ising)' city. The City had moved from the confidence afforded by empire to a more competitive and at times precarious status constituted out of new global and regional alignments.

In this chapter, the first of four locality-based case studies, I consider this prolonged struggle over place as a means of exploring the circulation of imperial sensibilities in this more uncertain postimperial present. Robins (1991: 23) argues that 'Empire has long been at the heart of British culture and imagination'. Certainly in the City of London the idea of empire is not confined to the past. It is an active memory which inhabits the present in a variety of practices and traditions and which still works to constitute the future of the City. Place plays an important role in the way in which memories of empire remain active. For example, the efforts to preserve the historic built environment in the present are often also efforts to preserve buildings and city scenes which memorialise the might of empire. Imperial nostalgias are not tied just to preservationist interests. New build schemes, such as the ones proposed for Bank Junction, may also activate a memory of empire even while they display a seeming disregard for its built environment legacy. Furthermore, both heritage and new build schemes are active place-making events in which social and economic visions are articulated. Imperial nostalgias, then, work through place in a multiple register. They are present in schemes to preserve what was and also in visions of what might be. But they are also present in the often discordant resonance between such place-making events and economic and social orders, real and imagined. The heritage battle of Bank Junction is about how an activated past assists in the City's (and the nation's) adjustment to the loss of empire.

Homi Bhabha argues, in the context of the 'narrated nation', that identity is constituted both from romanticised pleasures of hearth – an inward placeness – and from the terror of the race/space of the Other – an outward globalness (1990b: 2). For Bhabha, the local and the global are not set apart, but seen as constantly soliciting one another. The double geography of the global/local is not simply a matter of the global reaching into the local, it is also a matter of the local needing that which is not local in order to constitute itself. The quest for a sense of identity is not simply a calm return to an autochthonous essence. It is always also about an 'experience of division' which may set such imaginative returns in motion but might also activate outward-looking hatreds (Carter *et al.* 1993: xi; Salecl 1993: 103). During the height of empire this division was arranged through a map of power and influence in which the Other of the imperial Self was safely distant. For Britain, the experience of division is no longer ordered in quite the same way as it once was. This is not simply to say that Britain is now full of once colonised Others (an issue I elaborate in Chapter 4) but it is also to note that in the 'new world order' Britain has forged new global and regional alignments. The clearest of these, and the one most pertinent to the politics of identity and place which operates in the contemporary City of London, is what might be thought of as Britain's postimperial return to Europe. In the contemporary City of London, imperial nostalgias cohabit with the imperative of a creating regional alliance with Europe.

DIFFERENCE GATHERED IN THE CITY OF LONDON

Derrida, in his relevant discussion of the constitution of Europe, sees a 'culture of oneself *as* a culture *of* the other, a culture of the double genitive and of the *difference to oneself*'. In Derrida's conception, identity is constituted out of moments where 'difference remains *gathered*' (1992: 10–11). In the City of London of the 1980s there was indeed a 'gathering of difference'. This was starkly, but by no means solely, evident in two distinct narratives of place which had begun to gain popular currency in relation to the City of London. On the one hand, the City had been singled out by commentators like Saskia Sassen (1991) as a paradigmatic 'global city' in a new urban regime which transcends parochial nation-based urban hierarchies. In so describing London, and in particular the finance-centred City, such an explanation rests on the internationalisation and expansion of the finance sector and the trading in services, as well as a re-patterning of foreign investment facilitated by deregulation and transformed communications technology. The City of London is a space given over to finance and business.[1] London, it seems, had successfully shifted from the global geography of empire to the global geography of 'transterritorial markets' (ibid.: 327).

On the other hand, the City of London of the 1980s became implicated in a new urban design movement which advocated a commitment to indigenous architectural forms and a domestic, village-like townscape. Prior to the recent, more sensational and personal figuring of the Prince of Wales in the media, his public profile was linked to more sedate interventions in planning and architecture. His views on architecture and planning were brought to public attention through a Victoria and Albert exhibition, a television documentary and a book, all entitled *A Vision of Britain* (HRH The Prince of Wales 1989). The Prince's architectural programme is framed as deferentially indigenous, bowing to the natural and organic character of British architecture and townscape. His 'ten commandments' of architectural

FIGURE 2 The modernist Mies van der Rohe scheme for Bank Junction was initially commissioned in the 1960s. The Mansion House Square scheme addressed the surrounding townscape in name only. It took Palumbo almost twenty more years to accumulate all of the development site. By this time conservation and townscape planning had become the 'common wisdom' of the local authority and the scheme met with fierce opposition. (Source: Lloyd et al. 1976)

FIGURE 3 The James Stirling and Michael Wilford scheme for Bank Junction adopted a scale and style inspired by surrounding buildings. The more diminutive nomenclature of the scheme, 'No. 1 Poultry', discursively slips this grand vision into an 'undisturbed' streetscape. This rendition of the scheme was produced by the architect but was repeatedly referred to by those who claimed the scheme had a 'militaristic' feel. The aircraft flying overhead, which probably meant to signify the globality of the City, were perhaps too reminiscent of the more sinister and destructive flight paths of the Luftwaffe.
(Source: James Stirling, Michael Wilford and Associates)

design are presented modestly and mystically as if pieces of folklore – 'non-expert' views excavated from an architectural wisdom which springs forth from the land (ibid.: 75–153). Wright (1991: 239) suggests that Prince Charles 'makes his nation sound like a land of white bushmen'. His visions are at once both local and national in their reverberation; celebrating the local in order to restore a certain architectural order to the Kingdom. The defence of the local, the community, acknowledges the rights of ordinary places and people but in the service of a uniquely British (some would say English) scene and nation. The principles of the Prince of Wales contain a certain anxiety about a loss of order and are as much about nostalgia for stability as they are about the less manageable possibilities of local empowerment. He draws upon a wide range of architectural examples to furnish his polemic, but the City of London, and particularly the postwar transformation of its Canaletto skyline of church spires, is central to his argument about the demise of vernacular urban design more generally (see Daniels 1993: 12–13).

MAKING MONUMENTS

The thirty-year planning controversy surrounding the redevelopment of this site on Bank Junction has the distinction of being the longest running planning battle in London's history. Since the early 1960s the developer Peter Palumbo, through his property development firm, City Acre Property Trust, had been acquiring a group of Victorian buildings on Bank Junction with a view to placing on the site a prestigious office development of the highest architectural quality.[2] He presented two main schemes. The first scheme, commissioned in 1962 but not presented for formal planning approval until 1982, was an eighteen-storey modernist office tower designed by Ludwig Mies van der Rohe and grandly titled the Mansion House Square scheme (Figure 2). The second scheme was a five-storey postmodern office development designed by James Stirling of James Stirling, Michael Wilford and Associates and called the No. 1 Poultry scheme (Figure 3). Both proposals required the demolition of a block of Victorian buildings, eight of which were listed as historic buildings, and all standing within a designated Conservation Area (Figures 4, 5 and 6). Operating through the protection provided for under listed building and Conservation Area legislation, the local authority for the City, the Corporation of London, decided to refuse planning permission for both schemes on the grounds that they would seriously damage the historic character of the area. Using his statutory rights under existing planning legislation, Palumbo was able to challenge the local authority decision by asking the Secretary of State for the Environment to 'call in' the decisions for

FIGURES 4 and 5 (*over page*) The group of mainly nineteenth-century buildings on the Bank Junction development site were built as speculative office and retail space during the height of British imperialism. Although far from grand monuments, these buildings mark the more ordinary face of urban change produced by the invigorated trade and commerce of empire. The void left after demolition.

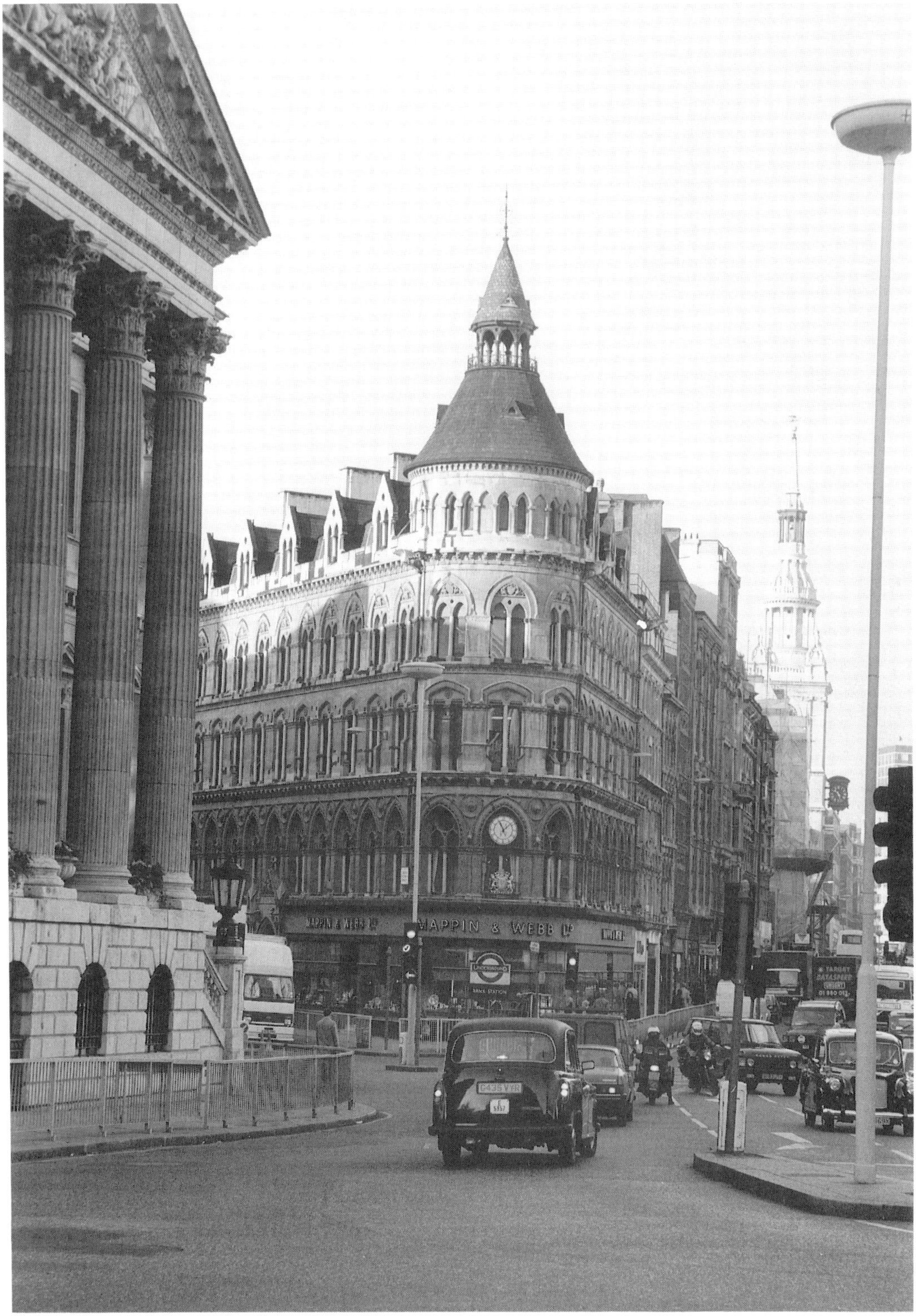
MAPPIN & WEBB Ld
MAPPIN & WEBB Ld

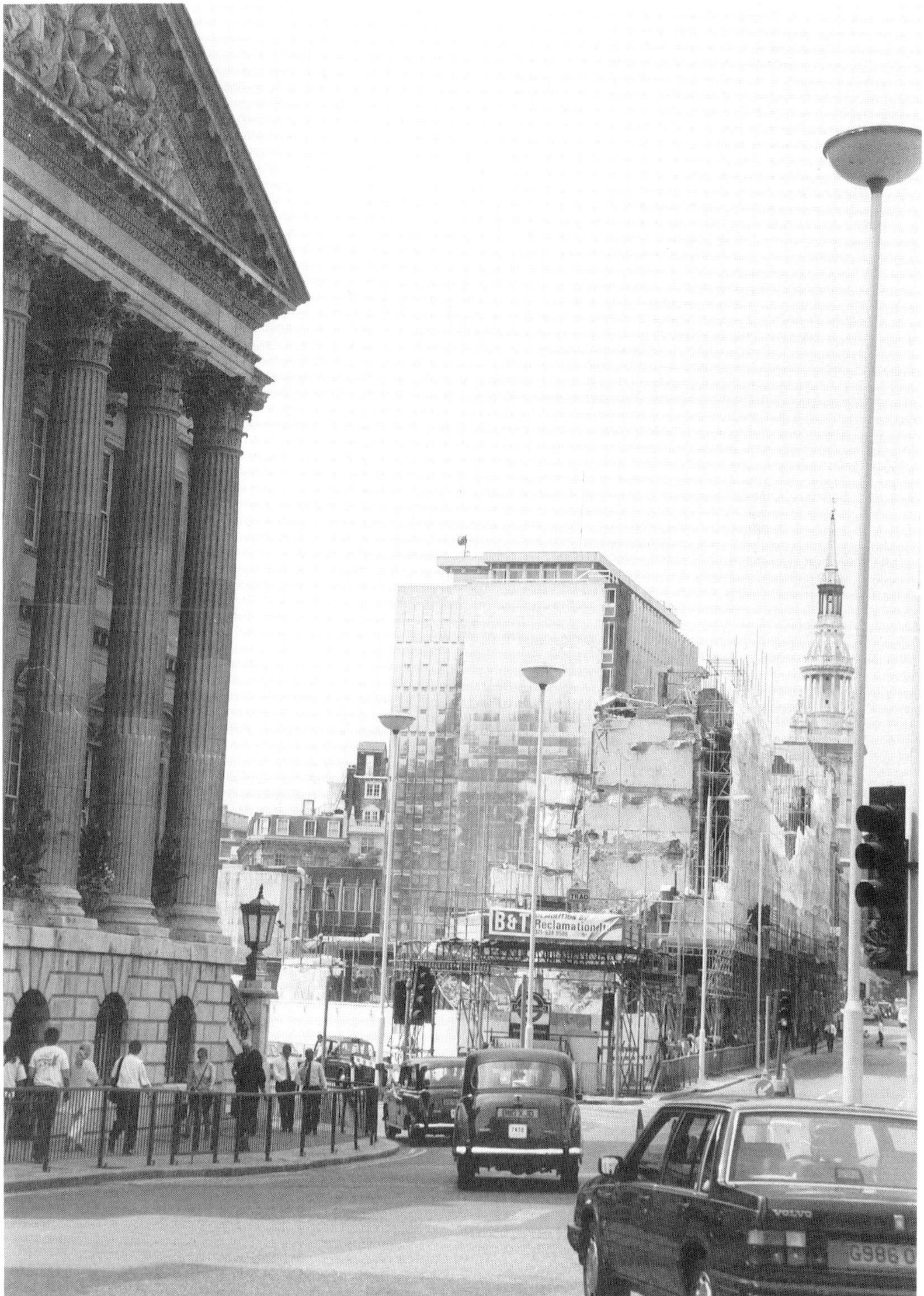
B&T
Reclamation
VOLVO
G986 O

FIGURE 6 The Bank Junction intersection is encrusted with grand, listed buildings: Soane's Bank of England, Dance's Mansion House, Tite's Royal Exchange and Lutyens's Midland Bank are the four most imposing buildings on the intersection. Slightly set back from the intersection is Hawksmoor's St Mary Woolnoth and Wren's St Stephen Walbrook which has a Henry Moore altarpiece donated by the developer Peter Palumbo. The area is entirely overlain with Conservation Area status which recognises the historical value of the individual buildings but also the townscape value of the buildings in relation to one another.

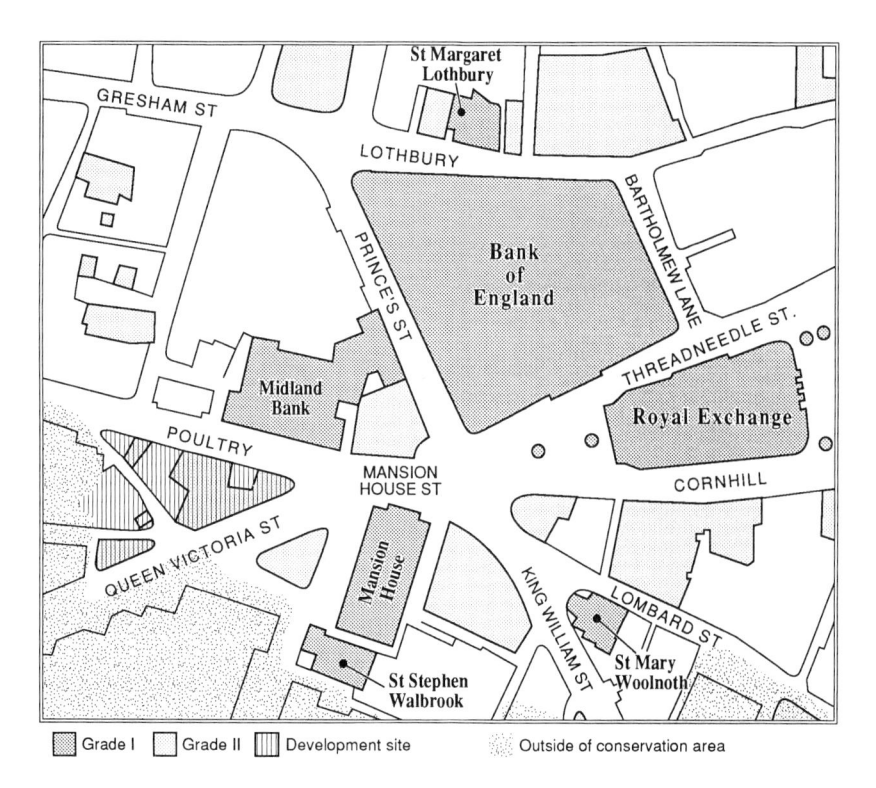

contestation in the arena of the public planning inquiry.

The proposals to redevelop part of Bank Junction went to public inquiry twice during the 1980s: the Mies tower in 1984 and the Stirling scheme in 1988. In the first inquiry the Mies tower was refused planning permission, but Palumbo was encouraged to submit a new scheme. The No. 1 Poultry scheme was commissioned but again went to a public inquiry. The findings of the second inquiry were against the granting of planning permission, but this recommendation was over-ruled by the Secretary of State for the Environment and planning permission was granted. The decision was challenged by conservationists and it was only through appeal to the Law Lords that permission to redevelop Bank Junction was finally secured.

The Corporation of London's refusals to grant planning permission were in part a response to obligations and options provided for under conservation legislation. Yet the Corporation's responses to Palumbo's visions were far from consistent. In the late 1960s, when Palumbo first mooted his vision for a modernist office tower on this central site, he was granted provisional planning permission. It was only when Palumbo applied for full planning permission some fifteen years later, after he had finally secured ownership of the majority of the development site, that he faced Corporation resistance in the form of the refusals to grant permission to build.

PICTURING THE EMPIRE

When Palumbo applied for full planning permission to redevelop his Bank Junction site in the mid-1980s, conservationists and the Corporation of London rallied to the defence of what heritage authorities had designated as 'a national architectural set piece' (English Heritage 1988). Those opposing demolition and new build development were, at one level, concerned with the loss of individual buildings that had historical merit and listed building status. But there was an equally strong concern for the collective value of the buildings on the development site and their relationship to the surrounding area. This was a concern for the townscape quality of the area and, in particular, the visual relationship between the more diminutive Victorian buildings on the development site and the surrounding cityscape. The Corporation of London engaged the services of townscape expert Roy Worskett to prepare and present its case for refusing planning permission to the scheme.

Since the 1970s the Corporation of London has vigorously pursued the conservation of the historic townscape quality of the City. In developing its conservation policy the Corporation's planning department was greatly influenced by a number of City townscape studies which were undertaken by experts drawn from the various historic building preservation lobby groups operating in London. The *Save the City* report designated a number of City precincts as worthy of special protection (Lloyd *et al*. 1976). These were later given planning legitimacy as designated Conservation Areas. The Conservation Area designations supplemented and shored up national listings of individual historic buildings which are heavily represented in the City core. By the mid-1980s approximately 70 per cent of the City was covered with designated Conservation Areas designed to protect the historic townscape qualities of the City (Corporation of London 1986). The 1986 Local Plan for the City was emphatic in its endorsement of the conservation/townscape approach to planning. The 'architecture, skyline and distinctive townscape' of the City were all to be 'preserved and enhanced' (Corporation of London 1986: 126). This has been implemented through restrictions on building heights, style guidelines, and the encouragement of refurbishment as opposed to demolition and new build.

The incorporation of the idea of 'townscape' into local City of London planning accorded with more general trends which shift the emphasis of conservation away from individual buildings. Townscape is concerned with the visual perception of the urban environment in compositional and pictorial terms: viewing cities as similar to paintings, 'as problems of composition, based on the production of a series of harmonies or contrasts ... the city as visual art' (R. Anderson 1988: 405). The key emphasis in such assessments is 'serial vision': the way in which elements of the urban scene interact visually as the observer moves through space. Townscape policy cherishes 'informality', 'accident' and 'spontaneity' but its creation and maintenance are contrived through active intervention in the urban scene, either through conservation or through the addition of certain built forms (Lowenthal and Prince 1965: 193). Townscape is now a commonsense notion in British and other planning systems.[3] Ironically, the translation of the townscape idea into urban policy has created a planning regime which regulates for acceptable visual 'disorder'.

Townscape as an approach to planning was initially developed and promoted by the editor of the British periodical *Architectural Review*, Hubert de Cronin Hastings.[4] He campaigned for a 'visual policy' of urban landscape, drawing on the eighteenth-century rural picturesque, which, in his view, was 'that landscaping tradition to which England owes its most personal aesthetic character' (de Cronin Hastings 1944: 5). The townscape concept was later given broader planning popularity through the writings of Gordon Cullen, one of the regular writers for the *Review*, who published a formal set of townscape principles (Cullen 1961). For Hastings, the English city was characterised by its 'infinite variety' and it was the task of planning to embolden 'irregularly' and 'disdain formality'. Hastings saw the responsibility of the planner to be the enhancement of inherited, 'natural', visual disorder – a state he dubbed '"sharawaggi", after an "East Asian" term for irregular gardening' (de Cronin Hastings 1944: 5).[5] This was an argument for the improvement of a 'scene according to the manner suggested by itself', a notion of development based on the *genius loci* of place, the intrinsic, indigenous qualities of the local.

The development and promotion of the townscape idea was set in direct contrast to modernist ideas of planning and architecture which had emerged on the continent in the inter-war years and had begun to appear in Britain. Townscape, Hastings argued, was more compatible with the English spirit and aesthetic and the English appreciation of 'age and quaintness' (de Cronin Hastings 1945: 165). Writing as 'I. de Wolfe' in 1949, Hastings proposed that this urban design policy would ensure that there was a uniquely English 'regional development of the International Style' (de Wolfe 1949: 355). Such ideas reflected Hastings's own combination of 'radical' Liberalism and nationalism – his commitment to individualism and the 'natural justice' of Britain's laissez-faire capitalism (de Cronin Hastings 1945: 167). The wartime context of Hastings's pronouncements on urban design was clearly relevant. The advocates of townscape sought to preserve and enhance the very Englishness threatened both physically and politically by the war. Townscape rejected the planning dogmatism of 1930s modernism, which itself had become emblematic of continental fascism (Esher 1983: 42). By confining urban interventions to the realm of the aesthetic it worked against comprehensive, state-led planning and the consequent threat this was seen as posing to the independence of private capital (Howells 1985: 29).

Matless (1990) has argued for the case of rural England, that processes of ordering and orchestrating 'indigenous disorder' are as deeply connected to the presentation of Englishness as the straightforward process of preserving what is old. In the controversy over the redevelopment of Bank Junction, the townscape assessments of the development site and its surrounds elaborated the status of this place as the symbolic heart of empire. And just as the idea of townscape emerged and has lived on through a tension with notions of 'foreign' continental European sensibilities of planning, so too did the assessments of townscape in this 1980s planning struggle express a domesticated memory of empire constructed in opposition to a demonised European other.

PLEASURES OF HEARTH

On the walk westwards along Cornhill into Bank Junction there is a short section of about fifteen paces where the dome of St Paul's looms in the skyline. In the Corporation's argument against the No. 1 Poultry redevelopment, this glimpsed view of the dome from Cornhill was claimed to be the 'most striking and significant aspect' of the Bank Junction area (Worskett 1988: 4). The No. 1 Poultry development proposal all but obliterates this glimpsed view, leaving only the cupola visible (Figure 7). In the Corporation's case, this existing view of the dome was considered nothing without the supplementary visual effect of the buildings of Bank Junction and particularly those on the appeal site. The Mappin and Webb turret of the existing Victorian buildings 'framed' and 'played' with the dome producing a 'superb kinetic view' (ibid.: 54).

The conservationist's defence of the glimpsed view of St Paul's dome from Cornhill is an extension of a long-held reverence for the visual supremacy of this great architectural piece of the City. St Paul's was the edifice of Wren's rebuilding of the City's churches after the Great Fire of 1666. Its status as a symbol of City survival gained new potency on the night of 29/30 December 1940, when the City faced one of its first direct German attacks of the Second World War. Almost one third of the City's fabric was destroyed in that and subsequent bombing raids. While the area immediately north, east and southeast was devastated by wartime bombing, the dome of St Paul's remained

FIGURE 7 The local view of St Paul's from the narrow street of Cornhill would be blocked by the proposed development. Opponents argued that this was a breach of existing local regulations to protect views of the dome. The developer argued that the No. 1 Poultry scheme would open out new views of the dome while the circular drum at the centre of the building would echo 'in absence' the proportions of the Cathedral dome. (Source: Roy Worskett 1989)

virtually intact. The image of the dome under attack from the Germans became a symbol of heroic British survival (Daniels 1993). Throughout the Corporation's postwar plans to remodel and rebuild the City there was continual reference to the need to protect and enhance the visual dominance of St Paul's. As early as 1934 the Corporation undertook its first study on height control in relation to St Paul's dome and restrictions became policy by 1935 (Kutcher 1976: 161).

Today, views of the dome of St Paul's, both from afar and from within the City, are marked and protected (Figure 8). The glimpse that might be had of the dome from Cornhill is the only view from the 'heart of the City' itself. For opponents of the No. 1 Poultry scheme this marked a special link between the eighteenth-century dome, the nineteenth-century city of empire and the present:

> [the view from Cornhill] is not just a view of St. Paul's from afar. It is the relationship between Bank Junction, Mansion House and the Mappin and Webb triangle and the metropolis and Empire.... [T]his viewpoint is ideal to give a sense of London as the economic centre of the Empire as well as the spiritual and other-worldly sense of the Empire.
>
> (English Heritage, No. 1 Poultry cross examination 1988)

Here the economic worldliness of empire, as embodied in the monuments of Bank Junction, is offered present-day moral absolution by the 'other-worldly' influences of St Paul's. Joseph Conrad wrote in *Heart of Darkness* (1902) that imperialism is 'not a pretty thing' and that its only redeeming feature is the 'idea at the back of it'. The battle of Bank Junction was also a struggle to preserve a spiritually redeemed heart of empire. In this struggle, the 'idea of empire' was once again taken away from its grimy past and placed into the foreground of a pretty scene in the present.

In the townscape argument against the No. 1 Poultry redevelopment, concern was also shown for the more immediate visual relationship between the existing group of Victorian buildings on the development site and other buildings abutting the Junction, such as the Bank of England, Mansion House and the Royal Exchange. The diversity and smallness of scale of the existing buildings on the development site were seen as relating positively to the monumentality of the surrounding buildings. The expert witness for the Corporation of London likened the relationship between the more humble Victorian buildings and the grander surrounding buildings to a 'theatrical show' in which the Victorian buildings were the 'supporting cast' to the 'stars' (Worskett 1988: 56). The 'visually subservient' nature of the buildings on the development site was seen to be their most important contribution to the character of the area and was likened to 'the relationship of visual master and servant' (ibid.: 39).

This recourse to the notion of urban hierarchy is a common feature of townscape rhetoric (Cullen 1961). It is presented as a benign ordering which provides visual diversity, a type of grammar necessary for the correct comprehension of urban form. However, the townscape concept is, like any other form of landscape idea, a social construction which naturalises the operations of power (see Daniels 1989). The hierarchical 'good manners' embodied in the principles of townscape reach beyond their immediate point of reference in the built environment of Bank Junction and address desired social orders. This broader resonance between the local townscape argument and an idealised social order is evident in what were then very public and quite influential views of the Prince of Wales on urban design. The Prince of Wales considers architectural hierarchy as especially important and includes it as one of his ten principles of planning. In advocating the need for urban hierarchy, the Prince makes a direct reference to the social order for which it is supposed to stand. In his view townscape hierarchy is common sense since 'civilized life is made more pleasurable by a shared understanding of simple rules of conduct' (HRH The Prince of Wales 1989: 80). Here one senses a nostalgia that extends beyond the heritage value of the built form, to a social and moral order once more surely held by the nation and reminiscently embodied in this symbolic site of empire.

EDGE OF EMPIRE

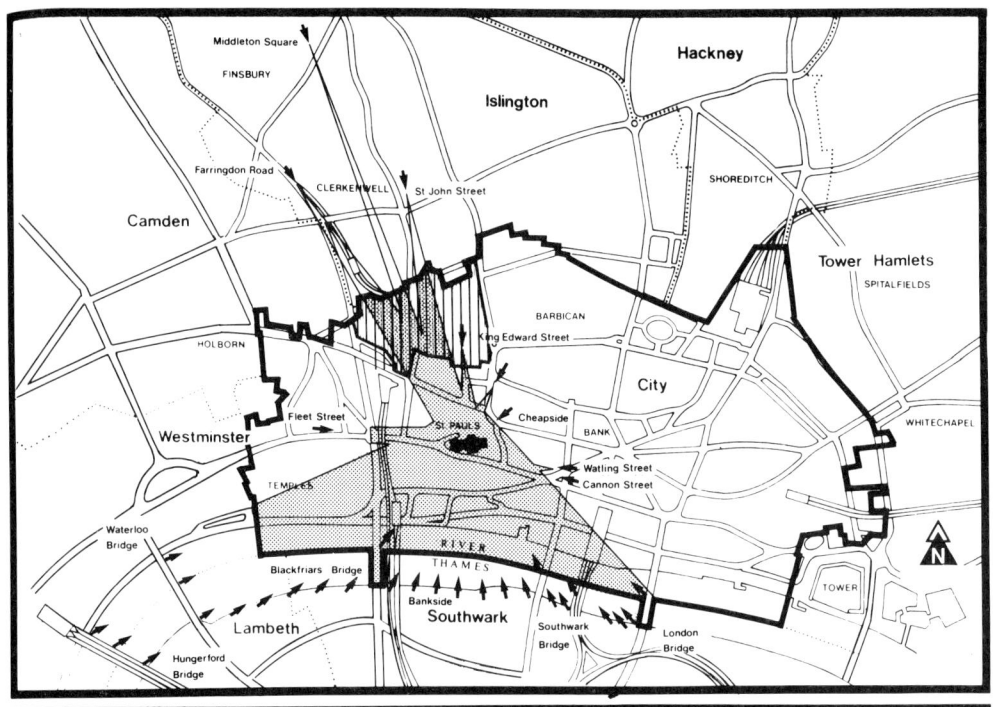

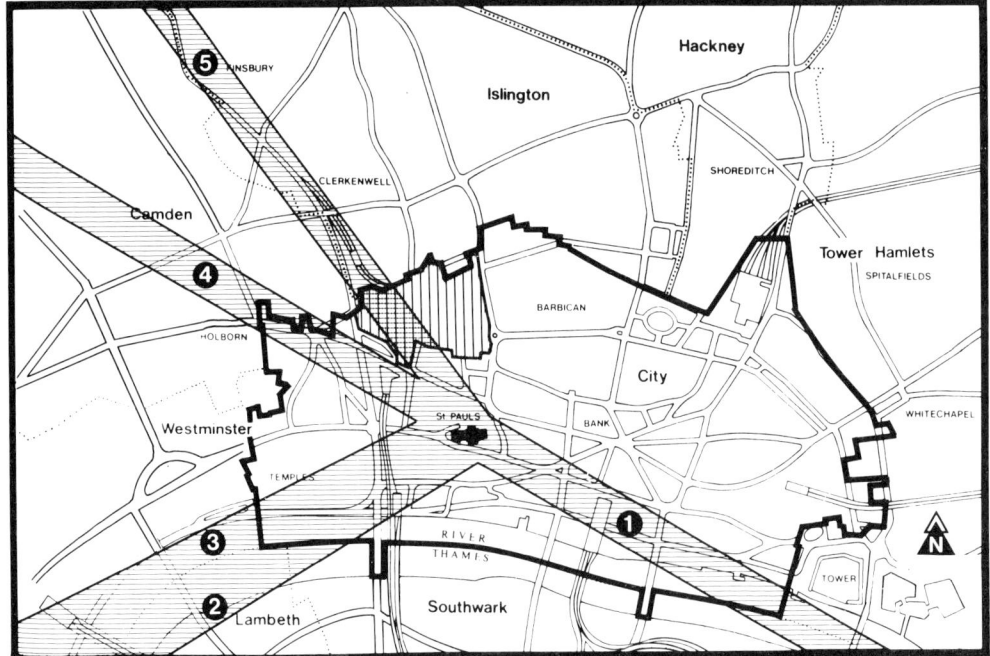

Height Controls

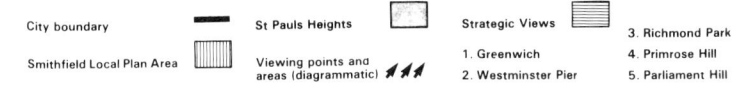

— City boundary
||||| Smithfield Local Plan Area
— St Pauls Heights
▨ Viewing points and areas (diagrammatic)
▦ Strategic Views

1. Greenwich
2. Westminster Pier
3. Richmond Park
4. Primrose Hill
5. Parliament Hill

Those contesting the proposed redevelopment turned to a particular image as incontrovertible evidence of the wisdom of their view. Hanging in majestic and nostalgic proof of evidence at the rear of the Public Inquiry room was Neils M. Lund's turn-of-the-century painting entitled *The Heart of the Empire* (see Figure 1). In this painting, Bank Junction is the visual hub of a city which seethes with the life of empire. Here is a place shot through with a vibrancy that had derived, quite literally, from the wealth obtained from exchanging raw materials from the colonies for manufactured goods in the core. The diminutive Victorian buildings of the development site play a crucial visual role in Lund's painting. They act as the visual pivot of the scene, an extension of the dome of St Paul's, guiding one's gaze to the dome and then beyond to the edge of empire. The battle of Bank Junction activated the memory of the City at a time when the politics of insurgent identities was safely at the empire's geographical margins: a memory of a time when the City of London felt secure as the centre of a global empire and where the social relations, both at home and abroad, were more surely set in its favour. Through the townscape argument the local planning authority and conservationists were constructing a symbolic terrain which spoke of values of morality, civility, hierarchy and order once central to the City of empire.

FIGURE 8 (*opposite*) The townscape logic of planning in the City of London was not simply concerned with local, intimate interplays between buildings of different scales. It reached its grandest expression in the various height control regulations which were passed to ensure that the dome of St Paul's Cathedral remained visible from within the City and from further afield. This concern with preserving the visual dominance of the dome of St Paul's was an important part of arguments both to protect the existing buildings on the development site and to rebuild the site. (Source: Corporation of London 1993c)

IMPERIAL ILLUSIONS

During the postwar years the City has shifted from an empire internationalism, with financial business tied to colonial modes of production and trade, to a new global internationalism (King 1990b: 9 and 83–87). The Euromarket (that is trading in currencies held off-shore from the country of origin) has been critically important in the transformation to a global city (Plender and Wallace 1985: 26; King 1990b: 91). For example, between 1965 and 1981 the size of the City's Euro-currency market had risen from US$11 billion to US$661 billion (Stafford 1992: 33). This market ensured that the City maintained and adjusted the source of its financial dominance in international terms. Not all sectors of the City enjoyed such expansion and in particular the operations of the Stock Exchange were seen to be hindering the City's capacity to compete effectively with other financial centres. Change came to the Stock Exchange in October 1986, in the form of the 'Big Bang'. Fixed commissions were abolished, allowing for single capacity trading, that is, brokers/dealers acting both as agents for others and on their own behalf in the buying and selling of stock (DEGW 1985: 8). The unlimited liability requirements, which had previously limited the companies that could join the Exchange, were also lifted. This 'deregulation' was accompanied by a major transformation in the technology and communications base of the financial sector. Screen-based trading was introduced which in some sectors has facilitated twenty-four-hour global trading. The impact of deregulation and the new technology on the financial sector was marked. Turnover in equities, for example, increased from an average of £650 million per day before the Big Bang to over £1.1 billion per day post-Big Bang (Clarke 1989: 125). The heady times following the Big Bang were tempered by the 1987 October Crash. The FTSE fell a record 250 points and it was estimated that some 3,000 City jobs were lost. Despite the Crash, such had been the impact of the internationalisation of banking and trading, deregulation and new technology, the City remained buoyant in its new-found 'global' status.

These various restructurings placed the pressure of change on existing social organisation and business practices within the City (Harris and Thane 1984; Budd and Whimster 1992). The City of London has always been simultaneously cosmopolitan and British (Cain and Hopkins 1993: 127). It has always accommodated difference but within the limits established by the class, race and gender-specific sociology of its financial practices: the stereotype of the pinstripe-suited, Oxbridge-educated, businessman operating through intricate structures of liveries and clubs (Lisle-Williams 1984; Cassis 1988). During the 1980s this surety was being shaken by a rapid phase of change which entailed new investment players, new financial institutions and conglomerates, new practices, and a changing and expanding labour force. For example, an emergent Euro-bond market in the City encouraged the growth of non-British banking interests. In 1914 there were only thirty foreign banks; by the 1930s this had expanded to over eighty and by the early 1960s there were over one hundred (Goodhart and Grant 1986: 9). From 1961 to 1971, the number of foreign banks in the City doubled and in the following decade doubled again. In 1987 there were 453 foreign banks either directly or indirectly represented in the City (King 1990b: 89). Thrift and Williams (1987) and Thrift (1994) have sketched the restructuring of the City labour force associated with the Big Bang and noted the ways in which it expanded rapidly and was associated with high pay and accelerated rates of salary growth. The managerial sector of the City's new workforce may have continued to conform to traditional class formations, but it is likely that the influx of younger workers to the City workforce transgressed familiar boundaries of class, race and gender (see also McDowell and Court 1994).

The restructuring of the City's traditional financial practices had a specific geographical expression. The new practices of the financial sector were increasingly dependent on computer-based trading and information technology and this generated demand for entirely new building types with different floor to ceiling heights to take cabling as well as large open internal layouts for conversational trading (Pryke 1991). There was also increased demand for high-quality office space in buildings with an architectural style that could be linked to the corporate identity of its occupants. This expanding and increasingly specialised demand for office space could not be met by existing City stock. In the six-month period following the Big Bang of 1986 the availability of office space fell by some 23 per cent (Richard Saunders and Partners 1986). Rents reflected the scarcity of suitable office space and in mid-1988 rents in the City centre had reached their all-time high of £60 per square foot, with the occasional rental of £70 per square foot. The way was open for a property development boom directed at meeting the new demand.

Palumbo argued for his development proposals on the basis of the global status of the City and this accelerating demand for quality office space. He may well have expected the local planning authority, the Corporation of London, to respond positively to such logic. The Corporation had not resisted the pressures for change in the City. While in 1982, for example, planning permission was given to just 689,000 square feet of office development, in 1987, the year following the Big Bang, the Corporation issued permits for over 12.5 million square feet of office floorspace either in the traditional core or on perimeter sites under its control (Corporation of London 1993a: 3). The mood of the local authority seemed to favour development. But the development ethos of the City had limits that were tied to the Corporation's views of how the City was best able to maintain its status as a 'global' centre.

The Corporation of London sought to differentiate itself from other competitors. The 1986 City of London plan elaborates:

> The City of London ... is noted for its business expertise, its wealth of history and its special architectural heritage. The combination of these three aspects gives the City a world-wide reputation which the Corporation is determined to foster and maintain.... The City's ambience is much valued and distinguishes it from other international business centres.
> (Corporation of London 1986: 3)

The preservation and enhancement of the local character of the City was seen as the very 'underpinning' of its global status and as an attraction to growth rather than as a deterrent. Heritage, correctly preserved and enhanced, was seen as the way the City could promote itself as distinctive in a new global market and allow it to compete effectively against challenges arising closer to home in the form of the Docklands redevelopments, the possible rise of

Frankfurt as a centre of a unified European financial community, as well as its 'global' competitors of Tokyo and New York.

The new communications technology associated with the 1980s restructuring of the City allowed an unprecedented spatial flexibility (Pryke 1991). Proximity to the core, which had long determined the spatial patterning of the City, was no longer necessary. Much of the post-Big Bang speculative development was located on or beyond the outer edges of the City, where development speculators could take advantage of large sites which had become available because of various closures and restructurings in the transport and manufacturing sectors (see Chapter 4). The second redevelopment scheme for Bank Junction offered a modest 125,000 square feet of lettable office space at a time when some 8 million square feet of City office construction was underway and planning permissions had been granted for a further 7.5 million square feet (Valuation Office, Inland Revenue 1988: 29). The Canary Wharf and other proposed Docklands Enterprise Zone developments added to this surfeit almost 12 million square feet of office space (London Planning Advisory Committee 1993: 45–46).

Without doubt, the conservation-mindedness of the local authority and its regulation of change to the built fabric of the City contributed to the diminished opportunities for development within the core areas of the square mile, so accelerating the City's outward push. New build development of the scale required by transformed City business practices was simply not available in the City core, which was almost entirely covered by designated Conservation Areas and where building up, as at the Canary Wharf development, was not possible within the strict height restrictions of local planning policy. As one spokesperson for the developer stated, a 'new business heart to the City' was being created on the border while 'the traditional heart is frozen as an historic monument' (Baker Harris Saunders 1988: 9). While it is easy to read the conservation tendencies of the local authority as a loyalty to the local vernacular, its effect may well have been to destabilise what it sought to preserve. It was not simply the appropriating forces of global capital (or the opportunities of new technology) that were decentring traditional City geographies. It was also the urge to preserve an historic built form in a context where new development opportunities were afforded by the opening up of land on the edge of the City.

By the time Palumbo's vision of the 1960s could begin to materialise in the 1980s and finally gain Law Lord approval in the early 1990s, it was not only his Mies van der Rohe office tower that seemed out of date, but his entire development strategy. This included his costly loyalty to the heart of the City as well as his unusual status as a lone developer in an urban centre under massive speculative development pressure from a new breed of property investment and development consortia (Healey 1990: 9; Pryke 1994). Furthermore, the last stage of Palumbo's struggle to redevelop this prime piece of real estate spanned a time when the property market went from boom to almost bust. By the late 1980s signs of oversupply were emerging, along with a contraction in property development activity. By 1989/90 there was some 9.58 million square feet of available office space in the City, an all-time high of 14 per cent of the stock, and it was speculated this would reach 17.5 per cent as new build developments reached completion (Valuation Office, Inland Revenue 1990: 47 and 1991: 49). The number of large units available for let in the City had also grown as many of the post-Big Bang schemes were completed. By 1989/90 there were forty-one office units over 51,000 square feet available for rent, with eighteen of these over 100,000 square feet. During this same period the take-up of office space was diminishing and by 1990 the take-up of lets was subject to massive incentives such as long rent-free periods.[6] During 1990/91 many of the larger property investment and development companies disappeared, including large players like Rosehaugh and Speyhawk, while rents for the central area of the City had retreated to a more modest £27–30 per square foot. It was in this climate of contraction that Palumbo's development proposal finally gained planning approval from the Parliamentary Law Lords. Not only had Palumbo been locked out of the boom period, he now had to actualise his dream in a contracting development phase.

The developer of Bank Junction and his opponents may have had different visions for the site but they shared a loyalty to this place as the symbolic heart of the City. This can be seen in the developer's shift from the uncompromisingly modernist Mies van der Rohe scheme to the postmodern style of James Stirling's No. 1 Poultry scheme. This was a strategic effort by Palumbo to adjust his development impulse to the architectural and planning sensibilities of the local setting in order to realise his development ambitions in the historicised inner core. The new No. 1 Poultry scheme conformed to local height restrictions. The development was largely confined to the existing street pattern. The design itself echoed the existing built form. Those advocating the No. 1 Poultry scheme argued that the design was inspired by local context and the historic character of the core. Stirling's design, his expert witnesses argued, was 'equal' to and 'did justice' to the surrounding monumental buildings of Bank Junction. While the scheme may have blocked the views of St Paul's from Cornhill, the central drum of the building echoed the dimensions of the dome and the tower had a viewing platform that would offer views of it never before seen. This design self-consciously sought to mimic all the qualities of the Junction which gave it its special place in the imperial nostalgias of the present.

Palumbo's deference to the local character of Bank Junction may well have been a clever appropriation of heritage by a developer desperate to see his vision materialise. But it is difficult to see this gladiatorial effort to develop Bank Junction as simply the strategic manoeuvrings of contemporary capital investment. Palumbo was driven by his desire to see a quality building of its time at the heart of a City which he, as much as his opponents, viewed as central to the memory of the grand and sure days of the City of empire. He was willing to wait thirty years, outlay massive professional fees and forego returns on his property investment in order to do so. And while the opposing sides were fully engaged in a 'battle', they each argued their case through the royally endorsed language and logic of townscape. They were, in their different ways, activating the memory of the heart of empire not through an outward-looking worldliness but through the vernacular of the local. This is the idea of empire brought back from its global reach and domesticated in the local.

CONTINENTAL ENTANGLEMENTS

The constitution of identity is not only marked by an inward turning to place. Of equal importance is the marking of difference through the notion of a feared Other. This marking of difference became the means by which the varying interests involved in the struggle to redevelop Bank Junction, actually unanimous in their loyalty to the importance of this place and the City, engaged in the performance of opposition. For Palumbo's opponents, the development schemes themselves were 'alien' to the local scene simply by virtue of being new buildings. But the full weight of this 'alien' status was elaborated by ascribing to the schemes broader social categories of Otherness.

Derrida reads a certain doubling in the constitution of Europe itself. It is either 'the point of departure for discovery, invention and colonization', or 'the middle, coiled up, indeed compressed along a Greco-Germanic axis, at the very center of the center of the cape' (Derrida 1992: 20). Britain now figures only faintly as part of the Europe of departure and colonisation. Its position in relation to the centre of the reconstituted European cape of economic union is also uncertain. In 1992, the year in which Europe struggled towards the finalisation of the Maastricht Treaty, a Eurosceptic writing in the Right-wing British magazine *The Spectator* proudly described Britain as 'a stroppy trading nation on the margins of Europe, ever striving to keep clear of continental entanglements' (Davies 1992: 12). Britain may well have been the pulsing hub of an empire, but the nation had an ambiguous and at times antagonistic relationship with continental Europe – particularly its Germanic core. The moves towards European economic and political union have meant that once again Britain faces its continental Other. This encounter between Britain and the Germanic 'center of the cape' manifested itself in the struggle to redevelop Bank Junction. Here the 'alien' nature of the proposed developments was elaborated through reinventing them as 'German' and thereby activating a familiar narrative of German antagonism towards Britain.

The first Palumbo proposal for Bank Junction, the modernist Mansion House Square scheme (see Figure 2), had an explicit German connection. The architect, Mies van der Rohe, was German-born. But the foreignness of Mies and his Mansion House Square scheme was not tied in a simple way to the architect's place of birth. In Mies's early years as an architect in Germany he fell out of favour with Hitler and fled to the United States. Indeed, it was not until the 1960s that Mies returned to Germany with a commission to design the National Gallery in Berlin. But in true International Style, the design used for this national monument was one originally commissioned for the unbuilt headquarters of Bacardi Rum in Santiago, Cuba (Balfour 1990: 224). Mies's design for Bank Junction was criticised by opponents not because of the architect's tenuous Germanic links but more because Mies, and his architecture, claimed to 'transcend the barriers of nationality' (Marks 1984: 39). As one of the advocates of the Mansion House Square scheme stressed, Mies was 'not an American architect, nor truly a German one, but an international architect in every sense' (Pawley, cited in ibid.: 49–50).

Mies's architecture is recognised as one of the purest forms of the 'International Style' to emerge from inter-war, continental Europe. Mies's vision, by any standards, had its own imperialist resonances. In his architectural philosophy it was possible to project an 'ideal space' on any place (Balfour 1990: 55). For Mies, the only 'real' architecture was that which 'touches the essence of the time' – which, for him, was not history but a transcendent order of geometrical reason (Mies van der Rohe, cited in Johnson 1978: 203–204). Opponents of the Mansion House Square scheme noted the 'curious' lack of any 'sense of place' in Mies's work. And the Corporation of London pointed to Mies's lack of familiarity with the City: he had visited the City only twice before presenting his scheme (Corporation of London, quoted in Marks 1984: 50–51).

It was this lack of sensitivity to the local that enabled Mies's tenuous Germanness to be reasserted and ascribed to the Mansion House Square scheme. The Prince of Wales's architectural vision provided the lead for this translation of a diminutive Little Englandism to a combative, xenophobic fear of the German Other. The Prince of Wales was outspoken about the Palumbo proposals for Bank Junction, labelling the Mies tower a 'glass stump' (HRH The Prince of Wales 1989: 66). More specifically, he likened the destruction that Palumbo's visions would wreak upon the City townscape as akin to that of the Luftwaffe (HRH The Prince of Wales 1987, quoted in Jencks 1988: 47).[7] Conservationists and architectural new traditionalists in Britain consistently earmark the Blitz as the beginning of the end for British architecture (Amery and Cruickshank 1975; Esher 1983). The wartime destruction of the City opened the way for massive reconstruction, much of which was executed in the International Style, with a clear lineage to the architects who emerged from 1930s continental Europe. This 'alien' architectural idiom is construed as much as an invasion of British soil as the Second World War air raids (HRH The Prince of Wales 1989: 9). In a publication advocating townscape planning, Tugnutt and Robertson (1987: 16) argue that modernist planning doctrines, 'largely imported from Europe', effected what they call a 'second Blitz'.

By 1988, when the No. 1 Poultry scheme was facing public planning inquiry, a further round in the political and economic restructuring of Europe was underway. The discursive constitution of Palumbo's second development scheme was set within a period marked by heightened British concern about European political and economic union and, in particular, fears of German ascendancy in the new Europe. The move towards Monetary Union has accentuated British concerns about German ascendancy in the new Europe. These are fears hard felt in the City where the specific anxiety is that it may relinquish its current dominance of the financial sector to Frankfurt, home of the Bundesbank, and a respectable financial centre in itself. A recent Corporation of London survey suggests that perceptions of European, specifically German ascendancy, are somewhat misplaced. In most of the major financial activities, the City remains in a strong position.[8] Yet the survey report notes that many City interests perceive Frankfurt to be 'by far the biggest threat' because it is actively pursuing some of the financial business traditionally centred in the City (Corporation of London 1994: 27). The Corporation report even makes the tellingly faint comparison that 'more people work in financial and business services in Greater London (over 700,000) than the entire population of Frankfurt (approximately 600,000)' (ibid.: 19). According to this report a worst-case scenario for the City would involve a single European currency which excluded sterling and which would be run by a German-based central bank (ibid.: 29).

Coakley (1992: 53) argues that British anxiety about economic union is in part registered around the perception of distinctly different economic models: the bank-dominated German model and the stock-market-dominated Anglo-Saxon model. It is fuelled by the knowledge that no matter how viable the City's financial and banking role, the economic bloc of Europe is dominated by the Deutschmark and, for the current moment, the Frankfurt-based Bundesbank. For example, the Conservative Government's resistance to

European Monetary Union was framed within a specific concern for the banks and security houses of the City which, it was argued, would find it difficult to participate in the financial markets, new market mechanisms and clearing and settlement systems associated with a single currency (Crawford 1993: 281). Already the path to a more elaborate European union has been far from smooth for Britain and this is perhaps most clearly demonstrated by the crisis over linking the pound sterling to the Exchange Rate Mechanism. The entry of the pound into the ERM was clouded by fears that this might mean a loss of monetary independence. Britain fully entered sterling into the ERM in 1990 with the pound set at an over-inflated rate of exchange supported by the government itself.[9] The will of the British Government to keep sterling afloat was not enough and it drifted towards its floor and, on 15 September 1992, fell below it when markets responded to a statement by the president of the Bundesbank that realignment was necessary and that this would require a number of currency devaluations (ibid.: 290). Sterling was floated out of the ERM until the speculative activity of the foreign exchange markets settled and the interest rates of Britain and Germany began to converge. Despite the fact that the fortunes of sterling during this period were clearly linked to local British political objectives, this crisis was widely read as a demonstration of the failure of the pound against the Mark at the hands of the Bundesbank.

It might be expected that the self-consciously 'local' design logic of the British (Scottish) architect James Stirling (see Figure 3) would evade charges of 'foreignness'. Yet Stirling's No. 1 Poultry scheme was not only cast as foreign but, in the context of the growing British 'crisis' with a restructuring Europe, its alien status was again ascribed by way of a specified 'Germanness' which drew upon the evocative war metaphor. The No. 1 Poultry scheme was described by opponents as 'aggressive', even 'militaristic' (Worskett 1988: 160–161). In the planning inquiry James Stirling was subjected to cross examination more like a wartime treason trial:

Corporation of London: You say that the prow does not overpower Mansion House, but is it not reminiscent of a German defence works?

James Stirling: No. I notice you refer not to English bunkers but to German ones.

Corporation of London: I am not saying German in a derogatory way . . . German bunkers are more powerful.

James Stirling: You obviously know German bunkers!
(Personal transcript, No. 1 Poultry Inquiry 1988)

FIGURE 9 The restrained deconstructive form and inventive play with classical architecture of James Stirling's Neue Staatsgalerie in Stuttgart. The architectural critic Charles Jencks lauded Stirling as one of the great postmodern architects of our time. Jencks was one of a star-studded line-up of witnesses to appear on behalf of the developer in the public planning inquiry into the No. 1 Poultry scheme. Palumbo is an avid patron of modern architecture and is reported to own houses designed by Mies, Lloyd Wright and Le Corbusier. (Source: Gardiner 1990: 22)

But it was actually James Stirling who was proven to have the more intimate knowledge of this building type. Under pressure of cross examination, Stirling 'confessed' that he had in fact been involved in modifying such defence bunkers (albeit British ones) after the war. Even the supposedly impartial public inquiry Inspector found it difficult to resist the appeal of this line of questioning. He brought to the inquiry a book on the buildings of the Channel Isle of Alderney, and showed a photograph from it of an 'Alderney Eyesore'. It was a German control tower, and legal advocates on both sides conceded that there was a striking similarity in style between this structure and the Stirling proposal (ibid.). Opponents of the development argued that while Stirling was acclaimed as one of Britain's 'big three' architects one of his most renowned buildings is the Neue Staatsgalerie in Stuttgart (Figure 9) and that some of his better known schemes in Britain were infamous for their design faults. Palumbo's opponents cast James Stirling as a 'traitor' who had an architectural style better suited to the taste and disposition of a demonised German Other. Palumbo's vision to redevelop was cast as an act of national subversion – an attack by the German core of Europe on the very heart of the City.

The intersection of James Stirling's architecture with a demonised Germany circulated more widely than the Bank Junction controversy itself. In 1988, Rover cars launched a new advertising campaign based on a narrative which featured the new Euro-businessman. In the television version of the advertisement two Euro-businessmen are depicted driving a stylish red Rover car through the streets of Stuttgart. The viewer knows what they say to each other because the brief 'dialogue' is subtitled, although subtitles are barely needed as this is a one-line script. The businessmen draw up in the Rover in front of the James Stirling-designed Neue Staatsgalerie. The passenger turns to the driver of the Rover and asks, 'Britischer Architekt?' This advertisement draws upon the direct competition that then existed between the German car manufacturer BMW and the British manufacturer Rover for a particular niche in the car market. In the Rover advertisement the British anxiety about German ascendancy in a new Europe is playfully reversed. Here German culture and capital are inhabited by British design. It was a reversal that was confined to the imaginative play of advertising, both in terms of the discursive constitution of Bank Junction and more generally. Gavin Stamp, one of the more vocal conservationists in the No. 1 Poultry case, took up the Rover advertising phrase 'Britischer Architekt' to title a damning appraisal of James Stirling's No. 1 Poultry scheme which appeared in *The Spectator* (Stamp 1988: 20). His appropriation of the Rover advertising slogan allowed him to articulate outside the confines of the planning inquiry the view that Stirling's architecture was alien to the City and symbolically linked to a narrative of German ascendancy over Britain. Neither the xenophobic opponents of James Stirling's design nor Rover's advertising agency could predict that in 1994 the latter's parent company, British Aerospace, would sell an 80 per cent controlling share in Rover holdings to BMW (*The Times* 1 February 1994).

The fate of Rover resonates with the fate of the Bank Junction development scheme. The lengthy and costly planning struggle Palumbo faced in seeing his vision materialise created an opening for the realisation of the fears of his opponents. Thirty years of development dreams, architectural schemes and legal costs meant that when final planning permission was received from the Law Lords, Palumbo needed to seek co-investors. Late in 1993 Palumbo secured a 50:50 joint venture partnership with Dieter Bock, chairman of the Frankfurt-based company Advanta AG. Bock's private partnership with Palumbo was to ensure that a scheme now estimated to have a development cost of £50 million would finally come to fruition. The construction of No. 1 Poultry is to be jointly financed by Advanta and Palumbo and, on completion, is to be jointly let and owned by the two parties (*Architects' Journal* 22 September 1993: 10). In one report readers were reminded that while the Prince of Wales likened the Stirling design to a giant '1930s wireless', the same scheme was not only financed, but also 'admired', by Dieter Bock (The *Independent* 22 September 1993). This final stage in the development saga may well speak to a more general trend in City property investment. A recent chartered surveyor's report based on all property deals in the City of London notes that German investment in City property is around 38 per cent compared with a UK investment of 39 per cent, the remainder given over to other foreign investors (Jones Lang and Wooton 1994: 4). Perhaps the feared 'invasion' is already underway and the beloved local of the City of London is 'inhabited', in a most material sense, by the Others of the global.

COLONIAL RETURNS

In April 1992 the IRA detonated a bomb in St Mary Axe near the Baltic Exchange, killing three people and causing considerable damage to nearby buildings, particularly the Commercial Union Building. A year later it bombed a site near the junction of Wormwood and Camomile Streets, badly damaging the London headquarters of the Hong Kong and Shanghai Banking Corporation and blasting windows from one of the City's tallest buildings, the NatWest Tower. The second bomb produced an overall damage bill of £1 billion, as well as killing one person and injuring forty-two others.

Northern Ireland is the oldest, most tenaciously monitored and, in these formally postcolonial times, decidedly idiosyncratic remnant of British imperialism. During the period in which formal British involvement moderated and, eventually, evaporated in most of the empire, intervention in Northern Ireland intensified. Attacks like those on the City of London were not a new feature of IRA activity and over the past few decades the IRA has frequently bombed sites in England including, of course, the notorious bombing of the Tory Party Conference in Brighton in 1984.

In choosing the City of London as its target in this new round of military activity the IRA was not simply attacking the symbolic heart of empire, it was also attacking the precariously placed Britain of the New Europe. The second bomb was timed to coincide with the annual meeting of the European Bank for Reconstruction and Development which attracted over 1,000 businessmen and bankers from around the world. One City banker remarked that the IRA 'could not have picked a better day to damage the City's reputation' (*Sunday Times* 25 April 1993). The media coverage of the bombings made much of the possible harm this spectacular display of the City's vulnerability would have on its chances of being chosen as the site of the European central bank. City interests were more stoic in their response and assured the world it was 'business as usual ... for the financial and commercial heart of Britain' (Corporation of London 1993d).

The Corporation of London's immediate response to the second bombing was tactlessly consistent with its over-riding conservationist agenda. The Planning Officer for the Corporation suggested that the damaging of the NatWest Tower provided the opportunity to demolish the building and replace it with something more suited to the City's character. As it is likely that both bombs were carried into the City in large vehicles, he also suggested that the bombings revealed the need for more rigid traffic restrictions in the City (*Sunday Telegraph* 2 May 1993). The bombings, which the Corporation acknowledged were a direct result of the City's role in financial activities, suggested to the Corporation that it accelerate its thinking on 'making the City less vulnerable as an economic target' and on making it a more 'livable' place which included residential space, pedestrian precincts and restricted traffic flows.

Although the IRA attacks on the City were conceived as acts of anti-colonialism, it was a most un-English City that suffered the direct force of the bombings. The full force of the second blast was most severely felt by the Hong Kong and Shanghai Banking Corporation, which was also home to the Saudi International Bank, the Bankodi Sicilia, the South African gold and platinum mining companies of Johannesburg Consolidated Investment and Barnato Bros, the Taai Bank and the Abu Dhabi Investment Authority. Also damaged were the National Bank of Abu Dhabi, the Banque Indosuez, the Deutsche Bank and the Long Term Credit Bank of Japan (*Guardian* 26 April 1993). Bombing a clearly English or British heart of empire was tough business in the City of London of the 1990s. Yet it is very likely that the bombing of these 'unhomely' targets had as much symbolic power as a bombing of the Bank of England, Mansion House or the Royal Exchange. It is precisely the presence of such international businesses that forms the foundation of the contemporary City and ensures that its diminished and domesticated idea of empire is translated into a new global context.

The bombings of the City of London set in train a range of emergency security operations which sought to assure local workers and international investors alike that the City was safe and secure. The first bombing in the City resulted in the implementation of 'Operation Rolling Rock' which increased the presence of uniformed police on the streets of the City and introduced random, short-term, road blocks at various City entry points. The second bombing encouraged the development of a long-term strategic plan to protect the heart of British financial and commercial activity. In the wake of the second bombing one senior City businessman suggested that the 'world's leading financial Capital' should erect a modern version of the medieval London Wall with steel security gates (*The Times* 27 April 1993). The Corporation of London argued against such heavy-handed security measures but, in conjunction with

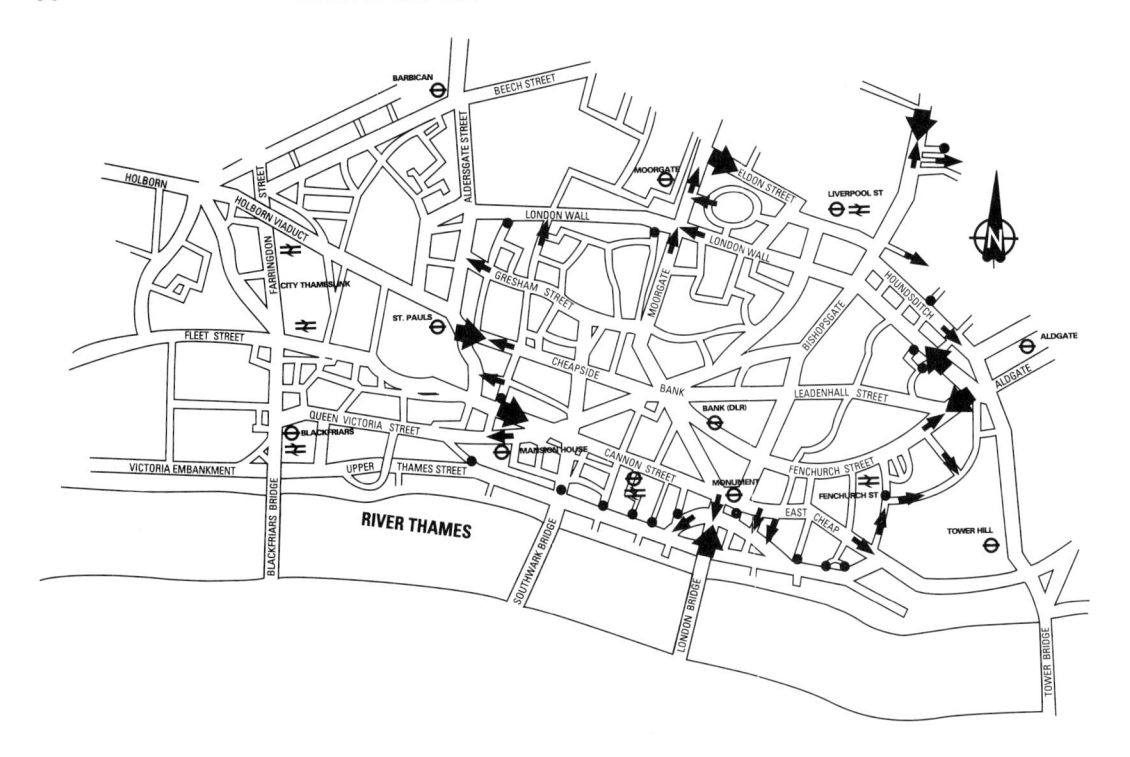

FIGURE 10 The Corporation of London responded to the IRA bombings by restricting traffic flow into and out of the City. The 'Ring of Steel' accorded with long-held conservationist aspirations to reduce the flow of vehicular traffic in the City. (Source: Corporation of London, 1993e)

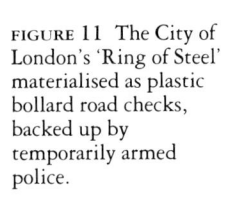

FIGURE 11 The City of London's 'Ring of Steel' materialised as plastic bollard road checks, backed up by temporarily armed police.

the City Police, did introduce armed police barricades at all the road entry points to the City (Figures 10 and 11). They called this security plan the 'Ring of Steel'. A second strategy called 'Camerawatch' encouraged private businesses to install their own video surveillance equipment to be linked to a central monitoring service. The aim of 'Camerawatch' was to ensure public areas within the City were monitored '24 hours a day 365 days a year' (City of London Police 1994: 2). Installation guidelines were issued by the Corporation to ensure that surveillance cameras were 'visually discreet' and did not adversely affect 'the character and general ambience of City streets' (Corporation of London 1993b).

In Mike Davis's (1990) analysis of the contemporary 'fortress city' he argues that the militarisation of urban life is a middle-class response to racialised or criminalised differentiations produced in cities by uneven economic and political restructurings. In Davis's Los Angeles, surveillance and walled spaces translate into a practice of social cleansing and segregation. The City of London's 'Ring of Steel' and its 'Camerawatch' surveillance system was a strategic response to a residual colonial predicament. It was specifically directed at keeping the IRA out of the City but not so that the City could return to some 'pure' space of the imperial imaginary. Rather, it was a strategy that guarded the City so that it might continue to negotiate its path towards the increasingly cosmopolitan requirements of being a 'global city'. Indeed the 're-walling' of the City of London sought to ensure a secure space for growth that would not, could not, be confined to the traditional geography of the square mile and was already spilling across these 'walls' into surrounding neighbourhoods.

Monuments may be formed from artefacts of the past or they may be made anew. In the planning saga of Bank Junction there is a struggle over how to make a monument to a City no longer clearly positioned at the centre; be it of a fading empire or an unpredictable global system. Although the protagonists clearly had different visions for the development site, they argued with a language and logic that were remarkably similar. No party in this struggle challenged the centrality of the Bank Junction to the City, and no party challenged the centrality of the City to the international status of the nation. In the end, the opponents were arguing simply about different ways to monumentalise the grandness of a place whose international status was under transformation and possibly threat. This planning 'battle' can actually be read as a thirty-year public ritual of reiteration and verification of a City of power. In this economic and imaginative transformation the idea of empire lived on and shaped the way the City moved into the future. The City of empire may well have constituted its 'pure' sense of Self from distant imperial relations. The City of the 1980s had to constitute its sense of Self, somewhat precariously, around a heartland that was full of the Others of the new global and regional order.

The redevelopment struggle of Bank Junction shows how the global and the local, the new and the old, the market and the vernacular, interacted and became the means by which the City's status and identity were renegotiated during a period of rapid change. It is tempting to equate 'the new' with less embedded, more superficial global forces of obliteration or, at best, an appropriative mimicry of the authentically local. Equally, it is tempting to align 'the old' with the diminutive, the local, a presumably more authentic embeddedness or possibly even resistance. The struggle of Bank Junction unsettled these alignments. Here local 'resistance' to change resonated with the reactionary nostalgias of royalty and a yearning for the purity of the idea of empire. Here 'change' itself became a nostalgic gesture towards a time when the City did more surely centre its geography around Bank Junction. Here a loyalty to the preservation and enhancement of the local built form actually worked to consolidate the decentring of historical geographies of power and the 'invasion' of the Heartland by a feared Other.

NOTES

1 During the 1980s, when the battle of Bank Junction raged, the City had a resident population of only 5,864 but a daily worker population of almost 300,000. Almost 30 per cent, the largest single proportion, of the workforce was involved in banking and finance (Corporation of London 1986).

2 Peter Palumbo was later knighted and given the task of heading the Arts Council.

3 The development of Conservation Area policy through the Civic Amenities Act 1967 gave formal planning status to the idea of townscape by establishing protective mechanisms for the 'cherished local scene' (Department of Environment, Circulars 23/77 and 8/87).

4 Hastings published his views in *The Architectural Review*, writing as The Editor or, occasionally, under the name of 'Ivor de Wolfe'. His views on townscape and the picturesque were reiterated in contributions to the journal by Nikolaus Pevsner and others.

5 Elaborating on the etymology of 'sharawaggi', Hastings notes the use of the term in English as early as 1685 when Sir William Temple was writing about the picturesque. Temple had attributed the term to the Chinese. Hastings also notes the use of the term in a 1940s publication by H. F. Clark on gardening and landscape. Clark attributed the term to the Japanese.

6   Take-up had become especially poor for older office stock and the practice of 'constructive vandalism', the deliberate stripping out of buildings to render them incapable of occupation and thereby reduce holding costs, became more common. The 697 property transactions recorded in 1991/92 was the lowest figure for a decade (Valuation Office, Inland Revenue 1992: 50).

7   This presented a paradoxical opposition for, as Wright (1991: 236) notes, many of the most avid critics of the Prince's architectural views saw his opposition to a modernist style to be as staunch as Hitler's.

8   The City is still the most active centre for international bank lending, although this is declining. It is still the largest world centre for trading in foreign equities, accounting for some 60 per cent of world turnover. It remains the largest centre for foreign exchange trading and still has a majority share of the international bond market (Corporation of London 1994).

9   Sterling was set at a central rate of DM 2.95 (and with upper and lower limits set at DM 3.132 and 2.778), a level which was chosen by the British Treasury with the inflation rate well in mind. This rate was read by economic commentators and the City alike as an over-evaluation. According to estimates in the financial press, intervention since the beginning of September 1990 to keep the base rates of sterling up had totalled some £30 billion including an obligatory intervention by the Bundesbank itself.

4

EASTERN TRADING

DIASPORAS, DWELLING AND PLACE

•

when there was a shape there was a reflection, and when there was a light there was a shadow, and when there was a sound there was an echo, and who could say where one had ended and the other had begun?

(Ackroyd 1985: 217)

In 1987 the Prince of Wales was reported to have gone on an 'East End safari' to the inner urban area of Spitalfields. He visited the area in his capacity as patron and initiator of Business in Community, one of the many new-style private/public community development initiatives now operating in Britain. Spitalfields is the economic and social Other to the City of London. Here wealth is not generated by trading in stocks and shares but often more precarious livings are made from small-scale garment manufacturing, retailing or restaurants. Spitalfields is the most deprived ward in Tower Hamlets, the most deprived borough in London. The area has some of the highest levels of housing stress in Britain. Over 80 per cent of residents live in public housing, and there is a persistent problem of mismatch between existing housing types and family size, leading to overcrowding. The unemployment rate of 17 per cent is well over that of the national average. At least 61 per cent (5,379) of the Spitalfields population is Bengali (Erens 1993), although some estimate the proportion to be as high as 80 per cent (London Research Centre figures, quoted in Community Development Group 1989: 19). By whatever measure, Spitalfields has one of the highest concentrations of Bengali people in Britain. They are the most recent in a succession of refugee and immigrant groups to have settled in Spitalfields, following on from the Huguenots in the eighteenth century, the Irish in the mid-nineteenth century and, later that century, Polish and Russian Jews. In 1987, when Prince Charles set out to tour one of his nation-making initiatives he encountered a place that he thought was less like home and, by his own words, more 'like a third world country' (*Guardian* 2 July 1987).

During the height of nineteenth-century imperialism, Britain's colonised

Other resided a safe distance from the imperial heartland. Geography was set in favour of preserving neat imperial divides between Self and Other, although both economically and imaginatively the British nation-home was forged through its dominions. The period of formal decolonisation and the postwar migrations of colonised peoples into British-based labour markets has brought the Other of the imperial Self back 'home'. As Stuart Hall (1991a: 24) remarks: 'As they hauled down the flag, we got on the banana boat and sailed right into London ... to the centre of the hub of the world.' Spitalfields may register as a place that is, formally speaking, a product of decolonisation; it might even be thought by some to be postcolonial. Yet for those who make the move from the former periphery to the former core, the postcoloniality of this reversal is likely to be less definitely felt. The logic of such movements – which groups migrate and where they migrate – is structured around the very specific relations of power, labour exploitation and obligation generated by imperialism. That is, although those migrating to Britain in the postwar/post-independence era may have been seen as 'outsiders', they were, 'moving within a system which already included them' (A. M. Smith 1994: 135). The outcomes of this movement, such as the conditions under which migrant groups live and their ability to register their interests and needs in their new home, are likely to be far from a state that might be thought of as postcolonial.

These resettlements have transgressed the frontier between core and periphery, creating 'immediate and intense' encounters (Robins 1991: 25). This proximity has activated an unstable renegotiation of imperialist arrangements of power, difference and distance. Such negotiations were always a feature of the daily execution of colonisation in distant lands, but more often than not the imperial heartland was artificially buffered from these unstable effects. Now that the Other of the colonial project is located within Britain's very shores it can no longer be placed as a known, but safely distant, supplement to the constitution of the nation. The postwar resettlements have produced a specific politics of race and nation. Indeed, Anna Marie Smith's (1994) deconstructive reading of postwar race policy in Britain shows that precisely at the moment when a corporeal edge of the empire came into the heart, politicians like Enoch Powell began the process of displacing a narrative of nation based on imperial might and replacing it with a narrative of nation based on an indigenous and pure Englishness. That is, the dependency of British power on the people and resources of the empire was defined as incidental and accidental, thereby rendering uncertain the right of the colonised to inhabit that heart, making them both 'outsiders', but also now 'an enemy within'. Powell's efforts to reorient the nation around itself at a time when that Self was confronting its supplementary Other as never before is one version of postwar British racism. In contemporary Britain such racism is articulated in many forms and is a daily part of the lives of migrant settler communities.

The processes of identity negotiation and destabilisation generated by the loss of empire and subsequent migrant settlements are clearly marked in contemporary Spitalfields. This chapter deals with two distinct, although not unrelated, types of urban transformation – gentrification and mega-scale redevelopment – and the politics of identity and place they activated. The process of the gentrification of the domestic housing stock of Spitalfields began in the 1970s. This gentler, domestic scale of change gave way, in the 1980s, to mega-scale redevelopment associated with the City of London making its speculative spill across its traditional boundaries. Large, abandoned and relatively cheap sites in Spitalfields made for easy and quick site assembly and economical, fast-track construction. At the end of the 1980s over 60 acres of land in the Spitalfields ward (some 20 per cent of its area) had been earmarked for large-scale development. In the vanguard of this speculative push eastwards from the City was the proposal to relocate the Spitalfields Wholesale Fruit and Vegetable Market and place on the 10-acre site a development intended to provide office space and services for the expanding City. Soon to follow were plans to redevelop the Bishopsgate Goodsyard and adjoining Truman's Brewery site (27 acres). These two developments alone proposed almost 2.5 million square feet of office space (and other uses) on an area totalling some 37 acres.

The co-presence of Bengali settlers, home-making gentrifiers and mega-scale developers activated an often conflictual politics of race and nation. Gilroy (1987: 288) argues that thinking through the politics of race is fundamental to understanding the 'production of urban meanings' and, as Keith and Cross (1993: 11) elaborate, to 'defining what city life is about'. The processes of urban transformation are part of the means by which a racialised architecture of power – material and ideological – operates. This is not simply a case of some 'imperialist' obliteration of the local by big capital. The colonial resonance of redevelopment lies in more than a convenient mirroring of imperialism's territorial expansions, frontier quests and 'foreign' invasions. Contemporary urban transformation is far more likely to engage consciously with the local character of an area than rapaciously obliterate it. This is perhaps most clearly seen in the varying ways in which heritage is mobilised as part of the legitimating framework of contemporary urban transformation. Through heritage designations and other architectural gestures towards the past, certain notions of place-based character are given material form and architectural precedence. Ideas of the Self and Other are unevenly freighted into the public domain and the material landscape through the making of heritage.

The struggles that emerged around these urban transformations produced a conflictual identity politics in which a variety of interest groups mobilised often contradictory notions of what Spitalfields stood for as a neighbourhood and what constituted the 'rightful' Spitalfields community. It produced an often antagonistic discursive terrain in which recalcitrant ideas of Englishness circulated, multicultural pasts and presents were marked, and new notions of being Bengali in Britain began to materialise. The conjuncture of Self/Other, core/periphery that occurred in Spitalfields by no means always resulted in the 'pure' racism that Enoch Powell signifies. The redevelopment of Spitalfields activated a politics of negotiation in which overt racism was one form of a range of competing processes of territorialisation which included more subtle processes of marginalisation and displacement (Goldberg 1993: 46). Yet this redevelopment phase also established some important opportunities for the Bengali settlers of Spitalfields to construct a homeplace in a new nation. This is an example, then, of the politics of how people – both British and Bengali – come to dwell 'in diasporic identities and heterogeneous histories' (Chambers 1994b: 246).

URBAN IMPERIALISMS

Metropolitan cities, like London, have long been inscribed with the language and logic of imperialism. This was not only registered in the grand monuments to imperial might as seen in, say, the City of London. Imperialist operations and ideologies also shaped urban spaces of deprivation. During the height of nineteenth-century imperial London, the East End docks handled goods traded between core and periphery and factories turned raw materials from the periphery into products for local and international consumption. The East End was a conduit in an imperial economy; the produce of the empire channelled through the East End but the wealth it generated came to rest elsewhere.

It was during the period of empire-dependent industrialisation and urbanisation that quite specific and persistent narratives of social nihilism began to adhere to the city. Speaking of 1844 England, Friedrich Engels (1971: 31) warned of a world split into its 'component atoms' and over a century later Herbert Marcuse (1964) was still certain that urban dwellers were 'one-dimensional men [sic]'. Both writers shared the belief, articulated by many other urban commentators, that under modernity there was a destructive colonisation of everyday life and a transformation from 'the manageable and safe Gemeinschaft' to the commodified and 'bureaucratized Gesellschaft' (I. M. Young 1990: 236; see also Gregory 1994: 403). This persistent sense of the ordeal of loss has produced a long search for a time and a place 'untouched' by modernity's 'kiss of death' (Berman 1982: 29). In *News from Nowhere* (1986 [1890]), for example, William Morris imagines a future that is a return to a pre-industrial time in which slums are transformed into village-like settlements modelled on a southern English rural scene. William Morris, of course, did more than just imagine less sinister, less urban, futures. His unique version of an English socialism gave rise to one of the first organised movements to seek to preserve the historic built form of the city and the nation, the Society for the Protection of Ancient Buildings (established in 1877). In William Morris's vision, the preservation of intrinsically English scenes and buildings was fundamental to the recovery of a 'nation's growth and hope': to ensuring that human relations triumphed over the outward-looking imperatives of economy, commerce and the state which accompanied imperial expansion (Morris, quoted in Thompson 1976: 228).

Places like the inner-city East End weaved in and out of the fabrication of a narrative of urban nihilism. The East End had a menacing presence in the nineteenth-century metropolitan imaginary. Neighbourhoods like Spitalfields provided Britain with irrefutable evidence of the casualties of the industrial expansion and mass urbanisation sustained by imperialism. The 'Darkest' East End acted as a localised Other against which the 'blaze of light', the imperial capital, constituted itself (R. Williams 1985: 229). Henry Mayhew constructed Eastenders as savage and exotic, while Charles Booth's survey of the East End systematically mapped and graded the poor, in an attempt to make this opaque but 'necessary evil' knowable and manageable (quoted in Raban 1988: 153). The apprehension that the East End produced in London's middle classes was placated only by the zealous efforts of various social reformers.

While nineteenth-century commentators held up the East End as a template of urban life gone wrong, mid-twentieth-century urbanists saw it as a site of residual hope. In the 1950s and 1960s, when British sociology was bent on rediscovering 'community' in the supposedly alienating metropolis, it was to the East End of London that they looked. The seminal work to emerge from this search was Young and Wilmott's *Family and Kinship in East London* (1957), but many others turned to the East End to verify the survival of community in the city through ethnographic documentation of the daily lives of the poor, white, working class. This recovery of a peaceable, white, working-class community may have been more than the nostalgia of urban modernity. These returns stand on the cusp of that moment when British cities were transformed by postwar migration from the colonies. They herald more sinister returns to whiteness as advocated by Enoch Powell's racist anti-migration rhetoric and policy. Indeed, in the Thatcherite period to follow, it was exactly the racialised inner city – areas like the East End – that became the focal point for revived discourses of 'urban crisis' and new articulations of imperialist administrative intervention (Goldberg 1993: 46). Instead of such areas providing the site of a return to a cosy community of whiteness, they marked the impossibility of such returns and, in the eyes of some, the possible destruction of the British nation.

Recent processes of inner-city renewal and transformation, such as gentrification and mega-scale redevelopment, have reversed the flow across the local frontier which divided East and West London (Schwarz 1991: 77). Within this reversal, brave (and astute) urban visionaries enter the territories of the poor inner city, to buy up cheap and supposedly 'unhomely' homes or large tracts of disused land and, through their restoration and redevelopment efforts, render them and the neighbourhood as a whole both more valuable and more 'civilised'. In the contemporary inner city, the idea of 'frontier' works to 'rationalise and legitimate a process of conquest' and 'urban pioneers, urban homesteaders, and urban cowboys are the new folk heroes' of this local imperialism (N. Smith 1986: 16–17). Through these urban 'renewals', 'productive' new worlds are carved out of 'unproductive' old ones.

LAND UNOCCUPIED

The modest eighteenth-century housing stock of Spitalfields has long been recognised by advocates of the protection of the historic built environment (Figure 12). The houses are architecturally distinctive by virtue of the studio mansards which were used by the early French Huguenot occupants to house their silk-weaving businesses (Survey of London 1957: 4–5). The first sustained conservation effort in Spitalfields came in the late 1950s when the area was chosen as a subject of study by the Survey of London, an organisation established to record systematically the historic built fabric of the city. According to the Survey, the deprivation, decay and 'evil reputation' of Spitalfields left the area virtually untouched by twentieth-century urban change and ensured the survival of the Georgian houses (ibid.: 1). In revalorising Spitalfields, deprivation and decay were displaced to allow a recovery of the precarious prosperity and 'good' society of the early French Huguenot weavers, known for their love of flowers and caged birds and their intellectual pursuits (ibid.: 7). The Georgian houses signified a more elegant, more prosperous, and more acceptably foreign Spitalfields. An emergent heritage aesthetic portrayed decay and neglect as 'benign' and allowed 'rotting brick' to speak not of 'social blight' but of 'a more radiant earlier age' (Wright 1987: 12). The architectural merit marked out by the Survey of London

FIGURE 12 The Georgian housing stock of Spitalfields was renowned for the distinctive mansard roofs where Huguenot weavers used to work. In this street the promiscuous mix of gentrified living and clothing workshop is clearly evident.

formed the basis of official recognition of the heritage value of Spitalfields. By 1969 the local planning authority for Spitalfields had designated three Conservation Areas and by 1976 most of the Georgian buildings in the area had been listed (Figure 13).

The Survey of London identified 230 eighteenth-century buildings in Spitalfields as having heritage value. By 1977 only 140 of these buildings remained, many lost to slum clearance or piecemeal demolitions (*Spitalfields Trust Newsletter* 1978: 1). This loss reactivated conservationist interests in Spitalfields during the late 1970s. British efforts to conserve the built environment can, by way of William Morris, lay claim to a socialist birth. But then, as now, this cause was largely the concern of 'a small knot of cultivated people' (May Morris, quoted in Thompson 1976: 241).[1] The re-emergence of conservation interest in Spitalfields was no exception. The efforts began in the mid-1970s in one of Britain's few remaining non-independent colonial outposts. A small coterie of friends of the Guinness heiress, the Marchioness of Dufferin and Ava, was enjoying a country-house weekend in Northern Ireland when conversation turned to the sad state of the Georgian housing stock in Spitalfields. It was decided that a charitable organisation called the Spitalfields Historic Buildings Trust would be established to save the area (Brien 1981: 6; Blain 1989: 6). One of the first delegations sent by the Trust to a prospective developer included, by the Trust's own admission, some of London's 'most eminent citizens' including the Features Editor of *Architects' Review*, the Chair of a major London publishing house, authors and television producers (*Spitalfields Trust Newsletter* 1977).

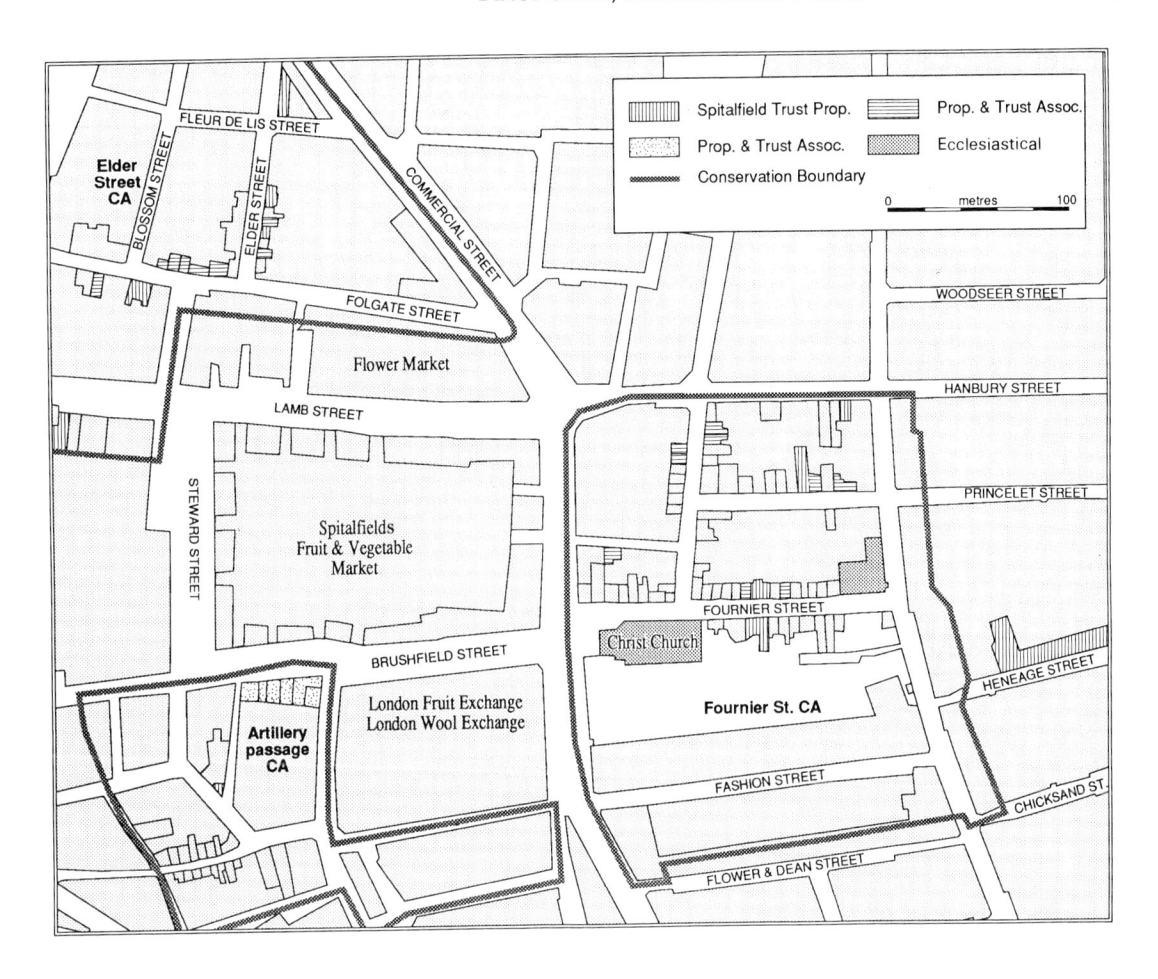

The Trust saw itself as more than a lobby group. It was an early example of a non-state, non-profit, community development trust and funded itself through a combination of public and private resources (Bailey 1990: 150–151). The Trust combined a form of guerrilla activism with business acumen. Its members squatted in threatened houses, formed delegations and publicised their efforts in the architectural and popular presses (Figure 14). A 'sleeping bag flying squad' acted by stealth, occupying threatened buildings the instant the enemy bulldozer appeared on the horizon. This was conceived as a 'battle' to 'save' what was seen as a derelict and neglected Spitalfields (Girouard *et al*.: 1989). The 'imperial' notion of an area bereft of life was a founding premise of the Trust's activity in Spitalfields. An early member of the Trust described his first impressions of the area:

FIGURE 13 The Spitalfields Market was surrounded by Conservation Areas and by houses that were listed on the official heritage register of English Heritage. Within this context the Market was seen as an alien and antagonistic land use.

FIGURE 14 The conservationist campaign to save Georgian houses from destruction and decay included direct action strategies such as squatting in threatened houses but also making good use of the press. The 'flying squad' kept vigilant look-out for houses which were under threat or for sale. The campaign brought together some unlikely types. Included in this group of squatters is the architectural writer Dan Cruickshank and Raphael Samuel, founding member of the History Workshop – a 'socialist history from below' collective. (Source: Spitalfields Historic Buildings Trust)

Staring across the glistening cobbles, all I could see in any direction was early Georgian domestic architecture. The scene was like an early photograph, taken on a long exposure and thus frozen in time, bereft of life: the subtle glow of purple and rose coloured brick, the glint of crown glass in every window, each house drab but dignified, untouched virtually, in all its details . . . grand silk merchant's houses, awe-inspiring despite shabby paintwork, broken railings and an air of general dereliction. . . . All those houses . . . were empty. Some of them had pathetic notices pinned up on their front doors, informing callers that such and such a business had closed down or moved elsewhere. Others were occupied by down and outs, having been deserted by their previous occupants following compulsory acquisition.

(Blain 1989: 2)

In nineteenth-century off-shore imperialism, heroics were rewarded by territorial possession. So too in 1970s Spitalfields. The Trust's forays into the inhospitable terrain of London's East End resulted in its triumphant possession of the jewels of Spitalfields: Georgian housing. The Trust established a specific programme of action to ensure private purchase of the Georgian houses. Some were purchased by the Trust itself, using donations or low-interest loans, secured through the connections members had with City-linked financial institutions and business dynasties. Between 1977 and 1987, the Trust purchased almost forty properties. Some of these were restored and re-leased, others sold on to appropriate buyers. The Trust also encouraged selected buyers to purchase houses on the private market. In the Trust's early years of operation, its Newsletter reads much like an estate agent's brochure, alerting prospective buyers to properties available for purchase and the government funds which could be used for restoration. The 'advertised' properties were lovingly described in terms of their particular Georgian features: the amount of original wood panelling present; the quality of the stair wells; the originality of the window plates; or the presence or not of a hallmark mansard roof. At its tenth anniversary celebrations in 1987 the Trust could boast that it had contributed directly or indirectly to the 'successful restoration of nearly 80 per cent of the early Georgian buildings' and the survival of 'this unique Georgian enclave' (*Spitalfields Trust Newsletter* 1987). As one founding member so aptly suggested, the Trust operated less like a conventional conservation group and more like an 'unofficial inner city development organisation' (Blain 1989: 9).

The Trust sought no ordinary home buyer. Those to be offered the Georgian houses, which were informally administered or officially owned by the Trust, had to be sympathetic to the organisation's ambition to restore Georgian Spitalfields carefully and faithfully to its eighteenth-century grandeur. The Trust regulated who moved into the area by controlled publicity. The early settlers of the reinvented Georgian Spitalfields either came by direct invitation, by word of mouth, or by reading about the houses in the Trust's Newsletter, which had a circulation limited to conservation, architecture and art circles (interview, Spitalfields Trust Member, 22 February 1989). To ensure that the new owners complied with the Trust's own, well-researched, vision of the Georgian aesthetic, it established a strict set of repair and restoration covenants and established a pool of reliable architects and craft workers 'in the William Morris mould' (*Spitalfields Trust Newsletter* 1979: 1).

HOGARTH AND SAG GOSHT

The 'saving of Spitalfields' relied upon the participation of financially and aesthetically equipped purchasers. This established the basis not only for the revalorisation of architectural Spitalfields, but also for the foundations of a new 'community' of gentrifiers. The Trust explained its social vision in its celebratory tenth anniversary history:

> a quite deliberate decision [was made], early in the Trust's life, to try and attract lively single people and couples who would make not only the maintenance of their homes but the revival of the area a high priority; in other words enthusiasts for whom a 1720s panelled townhouse would become more than just somewhere to live.
> (Blain 1989: 18)

The incoming population included home-working professionals (artists, architects and writers), residents of 'independent' means and included a notable proportion of childless households. The ethnic mix of incoming residents was not solely English, but was predominantly white European. The Trust kept close watch on the progress of this vision and their Newsletters reported enthusiastically and self-consciously on the number of Georgian houses to pass into the 'right' hands. Newsletter readers were informed about who moved in and their 'credentials', either as committed conservationists, sympathetic architects or adequately avant-garde artists and writers. The Trust strove for an architectural authenticity in its restoration work based on a strict aesthetic. Raphael Samuel, socialist historian, local resident and one-time member of the Trust, suggests that for all its emphasis on 'authenticity' the Trust has created an area that has 'an inescapable element of artifice' and where homes are transformed into 'showcases of the restorer's art' (Samuel 1989: 162–163).

The spectacle of restoration is most clearly expressed in one Spitalfields house which is both a dwelling and museum. Here the paying customer is taken back to the civilised everyday life of a family of Huguenot weavers. Taped conversations and sound effects are piped into each room. Presence is given to what is long gone by spilt glasses of wine and half-eaten fruit strewn across tables, clay pipes laying shattered in the fireplace and real urine lingering in an unemptied bed pot. A Hogarth print, which hangs above the fireplace, verifies the authenticity of the spectacle at hand. Here a familiar and cultivated French ethnicity folds into localised expressions of neo-Georgian Englishness.

Like many of the new community of Spitalfields, the creator of this museum/residence is openly proud of the social as well as architectural achievements of the Trust:

> It is one of the most fascinating groups of people I have ever come across anywhere.... We get along with the Bengali neighbours, everybody gets along, just because we are all so odd.... And as a community it is so busy producing things and doing things that it will go down a little like Bloomsbury, if we can get more produced. Because you know there are painters, writers, photographers ... it is very difficult to pick up a paper one day that doesn't somehow come near to one of us.
> (interview, Spitalfields resident, 14 April 1989)

In readiness for this future fame the museum/house includes an architectural memorial to the new community. The centrepiece of the dressing room is a Delft tiled fire surround. On first seeing the fireplace one imagines it is *c.* 1727, but on closer inspection one glimpses a most modern scene – two new Spitalfielders making love in a Porsche car parked in front of a lovingly restored Georgian terrace. The fire surround is *c.* 1981 and the tiles depict the new folk of Spitalfields (Figure 15). It is perhaps only right that this ardently nostalgic community memorialises itself before it has even passed. This ceramic collage of gentrified Spitalfields largely overlooks Jewish and Bengali revitalisations of the area. The historical period between the Huguenot occupation of Georgian Spitalfields and the contemporary neo-Georgian occupation is best depicted by a tile which shows a homeless man outside a house with a broken window and worn brickwork. This is the 'unoccupied' imaginary which preceded the 'saving' of Spitalfields by the Trust.

The creation of a social and architectural neo-Georgian enclave has had an

FIGURE 15 The Delft fire surround in one Spitalfields house self-consciously memorialised the gentrifier community. Included in the residents depicted were the London artists Gilbert and George.

enormous impact on Spitalfields. Properties that in 1978 could be bought for an average of £15,000, but in some cases as low as £3,500, were in the late 1980s selling for anything from £140,000 to £500,000 (Spitalfields Trust Records, cited in Forman 1989: 133). From its earliest days the Trust has always faced accusations of gentrification. These challenges came in part from the media which found it difficult to resist the temptation of reporting on some of London's better-known social figures taking up residence in this most unlikely part of London. The restored Spitalfields became the 'grandest slum' and home of the archetypal neo-Georgian (Artley and Robinson 1985: 13–15). There were also more localised accusations of gentrification which came from the local Left and various non-statutory voluntary sector organisations operating in the area. These loosely affiliated groups argued that the restoration of the housing stock to private home ownership displaced the Bengali-based garment manufacturing industry which often occupied the once affordable Georgian buildings.

The Trust claims it has never directly displaced any Bengali-based garment workshops. Indeed it engaged in a number of ventures designed to 'safeguard' Bengali businesses. As part of its grand design for the area, the Trust restored some 11,000 square feet of Victorian industrial building and associated properties for a garment manufacturing workshop, to be used specifically by those displaced from the Georgian housing stock or affected by the rising rents caused by gentrification. The provision of this alternative workshop space was heralded by the Trust as a necessary 'first step' in the 'far-reaching plan for the repair and the reconversion to residential use of much of the eighteenth-century housing stock ... without loss of jobs or undue disruption to the existing community' (*Spitalfields Trust Newsletter* 1979: 3). The Workshop project was regularly mobilised by the Trust as proof of its concern for the pre-existing social, cultural and economic community of Spitalfields and its commitment not to drive out 'locals' but to create a 'balanced' community. Yet the Trust's concerns over appropriate accommodation for Bengali workshops was also driven by their belief that these operations were having 'disastrous effect' on the 'fine but fragile' historic buildings of the area (ibid.).

It was into this local setting of incremental change and gentle antagonism that the mega-developments of the 1980s came. The first of these was the plan to relocate the Spitalfields Wholesale Fruit and Vegetable Market and develop the site for office, retail and other uses. The tender for the redevelopment was won by the Spitalfields Development Group (SDG).[2] The inflationary effect of the SDG redevelopment scheme was immediate. Under Market use the site was valued at £2 million. SDG took up its 150-year lease of the site for £60 million, with a ground rent of £500,000 per annum rising to 5 per cent of the rack rental value of the office element (Bernard Williams and Associates 1986: 7). In 1987 the total value of the proposed relocation and redevelopment scheme was put at £94,644,500 (Spitalfields Development Group 1987).

Local planning authorities saw development of the Spitalfields Market site as an aid, rather than a threat, to conservation objectives. The spread of new City-oriented development to areas on the fringe of the City was generally encouraged by the Corporation of London. In a localised form of neo-imperial expansionism, areas once peripheral to the City were re-evaluated as available for occupation precisely because they could 'afford relief from any excessive commercial development that might threaten the City's historic character' (Corporation of London 1987: 9–10). The local authority of Tower Hamlets

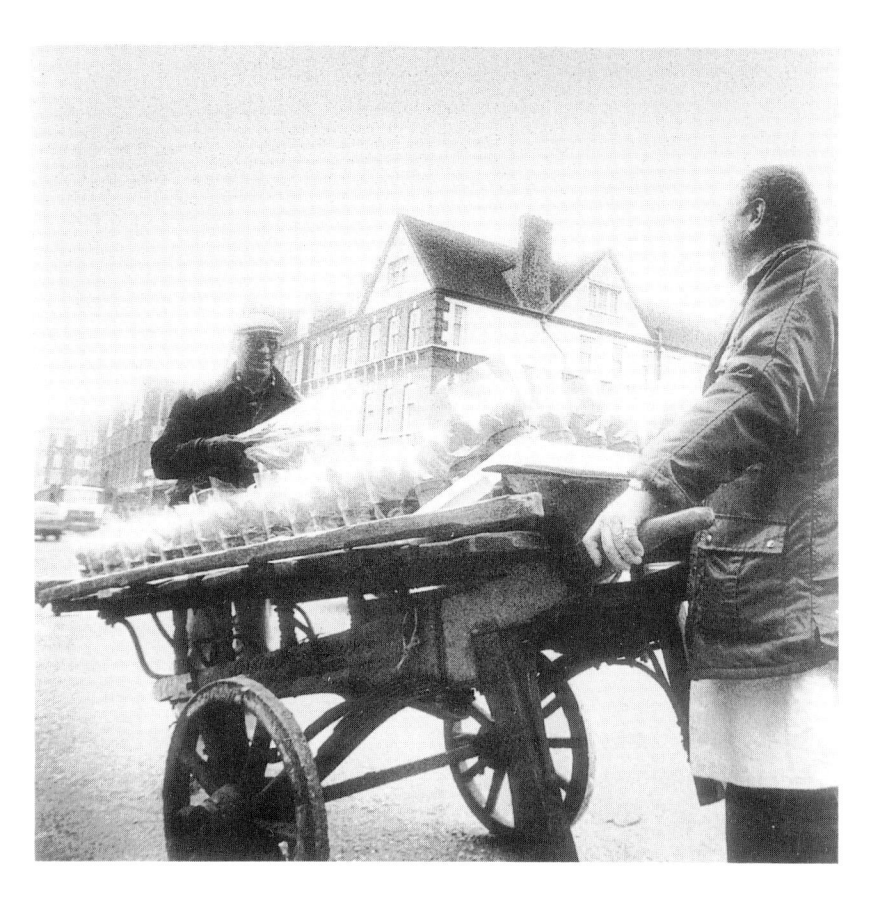

FIGURE 16 The Spitalfields Development Group produced a 'developer's history' of the area. The narrative proposed a continuous history of commerce and trade which culminated in the 'local logic' of the redevelopment scheme. The illustrations which accompanied the history showed romantic images of the daily life of Spitalfields. Here a Market trader wheels his wooden cart of produce through the deliberately ghosted streets of Spitalfields. (Source: Spitalfields Development Group)

saw the Market development as an opportunity to 'mend' the historic fabric of the area and create 'a prosperous living and working environment' (London Borough of Tower Hamlets 1986).

Stallybrass and White (1986: 27) conceptualise the traditional marketplace as both 'the epitome of local identity' and 'the unsettling of that identity by the trade and traffic of goods from elsewhere'. It is a place where there is an inevitable 'comingling of categories ... centre and periphery, inside and outside, stranger and local, commerce and festivity, high and low'. The Spitalfields Market, even in its most recent incarnation as a large-scale operation, certainly spoke of such comingling. The cosy produce-laden handbarrows and shouting costermongers (Figure 16) operated alongside rumbling juggernauts bearing produce from Europe and beyond. Initially the Trust viewed the 300-year-old Market as adding 'local colour' to the area and providing a link with the Spitalfields of old (*Spitalfields Trust Newsletter* 1980: 1). But as the Trust's Georgian Spitalfields became increasingly fixed in the built and social fabric of the area its tolerance of the Market as a source of 'local colour' diminished. The Trust's growing disdain towards the Market was clearly stated when redevelopment of the Market site was proposed.

In the controversy over the redevelopment of Bank Junction, discussed in the previous chapter, conservationists were set, in a familiar arrangement, against development. In the history of the plans to redevelop the Spitalfields Market site, conservationist opposition was unevenly expressed and, for much of the time, actively supported redevelopment, if not the architectural particulars of the schemes proposed. The Trust saw the redevelopment of the market as an opportunity to rid the area of elements which contradicted its vision to transform Spitalfields into a restored monument to early Georgian London. The rubbish generated by the Market operations, it argued, contributed to a 'feeling of decay' in the area while the traffic congestion and noise created by untimely night trading was 'a constant source of disturbance' (*Spitalfields Trust Newsletter* 1986: 2). The Market constantly challenged the pristinely restored Georgian Spitalfields, surrounding it with 'piles of rotting cabbages and tangled plastic packaging' (ibid.: 3). Systematic gentrification had imaginatively transformed Spitalfields into a 'clean' space and the Market into a 'dirty' space (Douglas 1984: 35). In neo-Georgian Spitalfields the double signification of the Market as both homely and unhomely was set asunder. The Trust was happy to see the Market replaced with a development which it felt 'worthy of this important and historic site' and which might provide for an improved, more wholly 'Georgian', Spitalfields.

To ensure the development contributed positively to the area, the Trust actively sought to influence its style and form. In conjunction with the locally based Georgian Group, the Trust produced a list of guidelines for the local authority's development brief and for the developers (Spitalfields Trust 1986a: 1). The guidelines included advice about traffic rerouting, the appropriate scale, building materials and street furnishings as well as specific recommendations to incorporate a group of listed buildings currently part of the Market and to model the overall plan on a Georgian street and square pattern. Through these guidelines, the conservationists introduced a range of design values which conformed with and elaborated their vision of a reconstituted Georgian Spitalfields (Figure 17).

The proposed redevelopment of the Wholesale Fruit and Vegetable Market also offered an opportunity for the Trust to further consolidate the spatial segregation between the white, middle- and upper-class homesteaders and the predominantly Bengali garment industry. The Trust suggested the redevelopment include workshops for the Bengali garment industry, arguing that the 'deployment of all remaining sweatshop workers to other premises' would safeguard their 'traditional' livelihoods (ibid.). It was a gesture entirely consistent with the Trust's objective of segregating residential and industrial uses in the area and the repatriation of the Georgian buildings to private

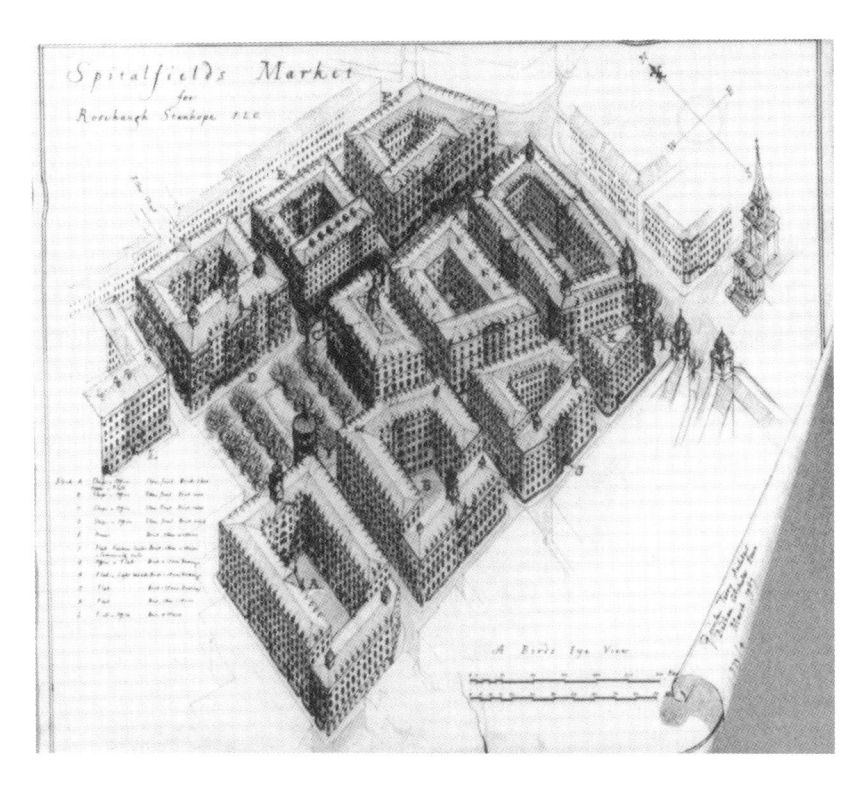

FIGURE 17 One of the many schemes which circulated in the tendering phase of the Spitalfields Market redevelopment was this overtly neo-classical scheme for Rosehaugh Stanhope. The architect, Quinlan Terry, is a well-known neo-classicist and his scheme was built upon a brief developed in close consultation with the local Georgian Group and the Prince of Wales's favoured architect/planner, Leon Krier. Quinlan Terry proposed an uncompromisingly classical scheme which was rigidly formal and disregarded the Grade II listed buildings in order to open out a diagonal vista to Hawksmoor's Christ Church. The scheme attempted to win the hearts of local conservationists but, ironically, they found it far too grand an appropriation of the Georgian architectural aesthetic. (Source: Jencks 1988)

residential use. Some twenty heritage listed houses in the areas were identified as continuing to house garment workshops which needed to be relocated.

The Spitalfields gentrifiers sought to reconstruct an 'authentic historical world' but this was most definitely a 'private' world with a preferred economic and social geography (Wright 1987: 14). The consumption practices of this coterie of urban specialists and artists created a new community in the inner city. Its nostalgic return to a restored Georgian Spitalfields produced an environment which was bathed in a rhetoric of cohabitation but was antagonistic to the Bengali occupation of the area. The ethnic origins of both Georgian and neo-Georgian Spitalfields were not exclusively English. Yet through the gentrifying imagination, Spitalfields came to stand as a domestic monument to the built heritage of the nation, a component of a sanctioned English heritage. Those aspects of Spitalfields that were incompatible with the conservation/gentrification agenda were at worst despised and always ultimately subject to the Trust's desires to see them spatially contained.

It is not simply that the Trust's activities directly or indirectly worked to squeeze Bengali garment workshops out of the Georgian houses and into more 'suitable' premises and places within Spitalfields. This desire has never been fully realised and even today at least some Bengali workshops remain cheek by jowl with gentrified houses. What is important in this history of genteel gentrification is the way in which an immediate encounter with Bengali Britons generated not an overt 'geography of rejection' (Sibley 1988: 410) but what might be better described as a managed 'multicultural' cohabitation. The Trust's aspirations for Spitalfields always engaged with the history of migration that characterised the area: the Huguenot, Jewish, Irish and now Bengali settlements. The Trust did not deny difference but regularly celebrated the 'traditional' mix of activities and peoples associated with Spitalfields. Indeed the Trust was instrumental in the purchase of No. 19 Princelet Street, a Georgian house converted into a Jewish synagogue, and in proposing that the building be turned into a 'Centre for the Study of Ethnic Minorities'. The research centre and museum is intended to pay homage to the diverse immigration history of Spitalfields. Recently renamed 'The Heritage Centre Spitalfields', the Centre is promoted as 'a new venture in the search for tolerance and understanding' (Spitalfields Heritage Centre 1987). Yet the Trust's own activities suggest that such 'knowing' cohabitation may well remain confined to the display spaces of museums. Its notion of a 'balanced community' was based upon a multiculturalism of convenience which provided a place (a discrete space) for Bengalis so that there might be more room for the elaboration of English heritage. The Trust's community vision did not generate a happy co-existence, rather it produced and legitimated development that was defensive and exclusionary (I. M. Young 1990: 235). The Trust's aspirations for the area were a search for an ordered new Britain in which a form of Little Englandism could exist untouched by the inescapable co-existence of a more disruptive postcolonial Otherness. If there was to be a multicultural Spitalfields, then it could not be unpredictably promiscuous, nor could it be comfortably Bengali. Being Bengali in the Trust's multicultural Spitalfields meant being spatially regulated, controlled along existing vectors of power, and, even better, staged as the last wave of immigrants in a tradition of immigration which belonged not to the Bengalis themselves but to a tolerant Englishness.

DEVELOPING NOSTALGIAS

Although heralding massive change for the area and responding to new City-based demands for office space, the developers of the Spitalfields Market site consciously framed their plans within an aesthetic and rhetoric which were sensitive to a historicised local. The logo of SDG was modelled on the listed buildings on the Market site which, in accordance with conservationist requests, were to be included in the new development. The pitched roof and chimney stacks of the Horner Buildings provided a logo which spoke more of

FIGURE 18 The Spitalfields Development Group logo was based on the heritage-listed Horner Buildings which were to be kept in the proposed redevelopment of the Market site. Their domestic scale and 'cosy' heritage feel provided a gentle face to a development that was large in scale and included high-density commercial development. Here the logo appears on a hoarding built around the development site. Save Spitalfields from the Developer Campaigners went on late night rounds and pasted their own 'Stop the City' posters over the logo.

a village townscape than an expansionist City (Figure 18 and see Figure 16). A public relations booklet promoted the scheme as a part of the 'continuing story' of Spitalfields (Spitalfields Development Group 1988). Here, past times and present development aspirations were intermeshed by the common historical thread of commerce. Spitalfields, the developer's history asserted, 'is more than bricks and mortar; it is living commerce'. In this developer's history, the scenario of decline and redemption was replaced by an unfailing logic of commercial adaptation and reinvention. The promotional history asserted that after redevelopment:

> the people of Spitalfields will still be doing much the same things as before. Spitalfields will be somewhere to live, relax, be entertained and shop; a place for people to work and prosper. All traditional pursuits, watched over by the weavers' houses, medieval precincts and Dickensian alleys and, above all, Hawksmoor's Christ Church.
>
> (ibid.: 12)

That the development was to provide new life for the community (jobs, housing, shops) was a persistent theme in the developer's rhetoric. Planning provisions which allowed for 'planning gains' to be passed from developers to local communities ensured that in part the promotional imaginings of SDG had some material effect. Some £5 million was paid to a Community Trust to provide funds for local development, and an annual payment of £150,000 was to be paid into a training scheme for a five-year period. Guarantees were also made to include in the final scheme some 118 social housing units, a crèche and community centre as well as publicly accessible open space.³

The developers associated with the Market site were quick to appropriate the conservation-driven design aesthetic valorised through recent gentrification. The original scheme with which SDG won the Market tender was designed by architects Fitzroy Robinson in conjunction with design architect Richard MacCormac (Figure 19). MacCormac designed a diverse scheme within which a series of 'architectural conversations' could take place both between the different buildings on the site and the surrounding townscape (interview, Richard MacCormac, 15 June 1989). The scheme 'teased' with local architectural idioms and kept the listed Horner Buildings as a signature component. The site plan built upon a fine-grained network of streets, corridors, arcades and open spaces which imitated the surrounding Conservation Areas.

MacCormac was a logical choice as architect for developers who needed to secure planning approval as a necessary preliminary step to winning the development contract. He was a founding member of the Spitalfields Trust and since 1979 had lived and worked in Spitalfields in a nineteenth-century

FIGURE 19 The MacCormac scheme for the Spitalfields Market redevelopment. It attempted to create a soft edge between the surrounding gentrified neighbourhoods and the commercial development within. Richard MacCormac is a Spitalfields resident and considered to be one of the creative 'new community'. He is depicted on the community fire surround in Dennis Severs's self-conscious heritage house. (Source: Spitalfields Development Group 1987)

brewery which the Trust had 'rescued'. MacCormac had long expressed his own vision for redevelopment and revitalisation of the area (MacCormac 1983; Cruickshank 1989). His loyalty did not extend to the Spitalfields Market which he saw as one of the sources of the decline of Spitalfields: a 'foreign' element in the area which has worked to 'cut' Spitalfields out of the 'psychological geography' of London (interview, Richard MacCormac, 15 June 1989). The MacCormac scheme met with approval both from local planners and conservationists.

Since the early MacCormac scheme the Market redevelopment has been through a number of significant changes. The appointment of an American firm, Swanke Hayden Connell, as 'administrative' architects resulted in MacCormac withdrawing from the project and an entirely new scheme being proposed. The Swanke Hayden Connell plans increased office provision and the overall bulk of the scheme. This scheme met with fierce opposition from local and national conservationist interests. The Spitalfields Trust considered the new scheme to have 'no sense of history or even of style' and asked that the architects be 'sent back where they came from' (letter, Spitalfields Trust to SDG, 8 December 1989, in Spitalfields Trust Records). Lord St John Stevas of Fawsley, of the Royal Fine Art Commission, described the Swanke Hayden Connell scheme as entirely unsuitable; a plan by American architects to 'wreak revenge on their old colonial rulers by undermining our historic capital' (Lord St John Stevas, quoted in *Architects' Journal* 25 July 1990).

The Trust began to campaign against the redevelopment and asked that the scheme be called in by the Secretary of State for the Environment for public planning inquiry. After four years of collusion based on a shared discourse of architecture and urban design, in this new oppositional mode the very community the Trust had attempted to contain and even displace was mobilised in defence of the Trust's urban aesthetic.

> We are increasingly concerned about the effect of this mega-development on the life of the existing community, along with the more direct impact on the Conservation Areas.... [Allowing] office use is putting increased commercial pressure on traditional workshop spaces for the local rag trade.
> (*Spitalfields Trust Newsletter* 1989: 1)

In 1989, under the force of conservationist opposition, the Swanke Hayden Connell scheme was called in for public planning inquiry. The developers, wishing to avoid the costly public inquiry delays, withdrew their scheme and began to conjure yet another plan for the site under the guidance of US architect Ben Thompson, who is well known for his inner-city restoration and revitalisation projects. Anti-American sentiments were placated by the appointment of a British design and planning advisory panel and a stable of British architects to design specific phases of the scheme. The new development strategy adopted an incremental approach which effectively subdivided the site and allowed for piecemeal development. The large site, single use 'groundscrapers' of the 1980s (S. Williams 1992) were in the early 1990s being replaced with a more flexible development strategy better suited to negotiating both global economic shifts and local political vagaries.

Some seven years after the redevelopment was proposed, and numerous designs later, the vision remained unrealised. It was reported that the Market redevelopment was a financial disaster, having already cost SDG some £200 million. In 1991 SDG still owed the Corporation of London, owners of the Market site, some £50 million and faced a total development cost of £500 million (*Evening Standard* 17 July 1992). By August of 1992 SDG was reported to be discussing with banks the rollover of a £165-million debt.

TRADING IN COMMUNITY

While conservationists were engaged in a genteel reassertion of Englishness in the Georgian streets of Spitalfields, more rampantly nationalist and racist forces were elsewhere at work. Brick Lane is the artery of Spitalfields: it is lined with Bengali-run small businesses (clothing and leather garment shops, numerous restaurants and food shops), and contains the main mosque of the area, as well as various community service shopfronts. The main language of the street is Bengali and the air is heavy with the smell of curry. Since the 1970s Brick Lane has become the focal point for an ongoing and organised campaign of racism, generated by the Right-wing British National Party, an offshoot of the National Front (Figure 20). Large groups of BNP/National Front supporters collect in Brick Lane on weekend market days to sell their wares and noisily make their claims about the invasion of British soil by 'Asians'. Throughout the 1970s racial violence was commonplace in Brick Lane and its surrounds. In 1978 a young Bengali clothing worker, Altab Ali, was murdered. The event triggered a wave of anti-racist protests in the area, including a large, predominantly Bengali, march from Brick Lane to Downing Street (Figure 21). Anti-racist protests were met with increasingly systematic displays of racial harassment, including a rampage of youths through Brick Lane. In the various accounts of these events which emerged, including anti-racist accounts, Brick Lane became the geographical signifier of racial strife in the East End (Leech 1994: 15).

The racial strife produced by these reactionary groups fuelled the

FIGURE 20 Racial harassment is an everyday part of life for Bengali residents of Spitalfields. The National Front has since the 1970s consciously concentrated their efforts on the Brick Lane area. (Reproduced by courtesy of Paul Trevor)

emergent view under Thatcherism that Britain was in the grip of a 'crisis' which could be pinned down to the racialised and criminalised spaces of the inner city. Many of Thatcher's most strident restructurings and interventions focused on the inner city, although they were as much to do with breaking Labour's hold in these areas as calming an urban 'crisis'. In London, a number of traditional Left strongholds were destroyed, most notably the Greater London Council and the Inner London Education Authority. In Spitalfields, like many other inner-city areas, a range of smaller local agencies which had long worked in sympathy with the Left were systematically dismantled to be replaced with market-linked development strategies. These interventions produced a particular crisis for the Left, which had traditionally relied upon the 'life-belt' of constituents in the inner city (Mythen 1991: 90). In the 1986 local elections, Spitalfields became a lone Labour ward in a Neighbourhood Area dominated, for the first time, by the Liberals.

In recent years British socialism has moved towards more localised political practice focused on single-issue alliances within the community and extra-parliamentary activity (Gyford 1985). Some commentators have suggested that this turn to the local has meant that there is now 'a lot of ... "Englishness" about in Labour and socialist circles' (Yeo 1986: 311). But in many inner-city areas a local socialism has provided the basis for the Left to engage with a relatively new constituency: those racialised 'settlers' who have a more ambiguous positioning in terms of the traditional socialist goals of a universal class struggle. For the local Left of Spitalfields the issue of racial harassment provided an important focal point around which to build alliances with the Bengali community. Increasingly estranged from government, the task of protecting Spitalfields as a safe haven for minority groups became a feature of the local Left's extra-parliamentary political activity.

In the 1980s the attention of the local Left was redirected towards the proposal to relocate the Market and redevelop the site to meet the needs of City expansion (Figure 22). The redevelopment was tenaciously opposed through a group called the 'Campaign to Save Spitalfields from the Developer' (CSSD), which was an alliance between the local Labour Party and what remained of a decimated community service sector. The normal forum for such opposition is a formal public planning inquiry and the Campaign made numerous unsuccessful appeals for an inquiry. Its failure in this regard is in part due to the discursive parameters of current planning politics in Britain. Whether in collusion or in conflict, conservationists and developers share a common discursive terrain based on a privileging of urban design concerns and a shared commitment to the regenerative potential of private capital. Planning

FIGURE 21 During the 1970s and early 1980s a string of racial attacks occurred in the Spitalfields area. The death of one youth resulted not only in large protest marches but also in local politicisation of the Bengali community. During this period important political links were forged with the local Labour Party. (Reproduced by courtesy of Paul Trevor)

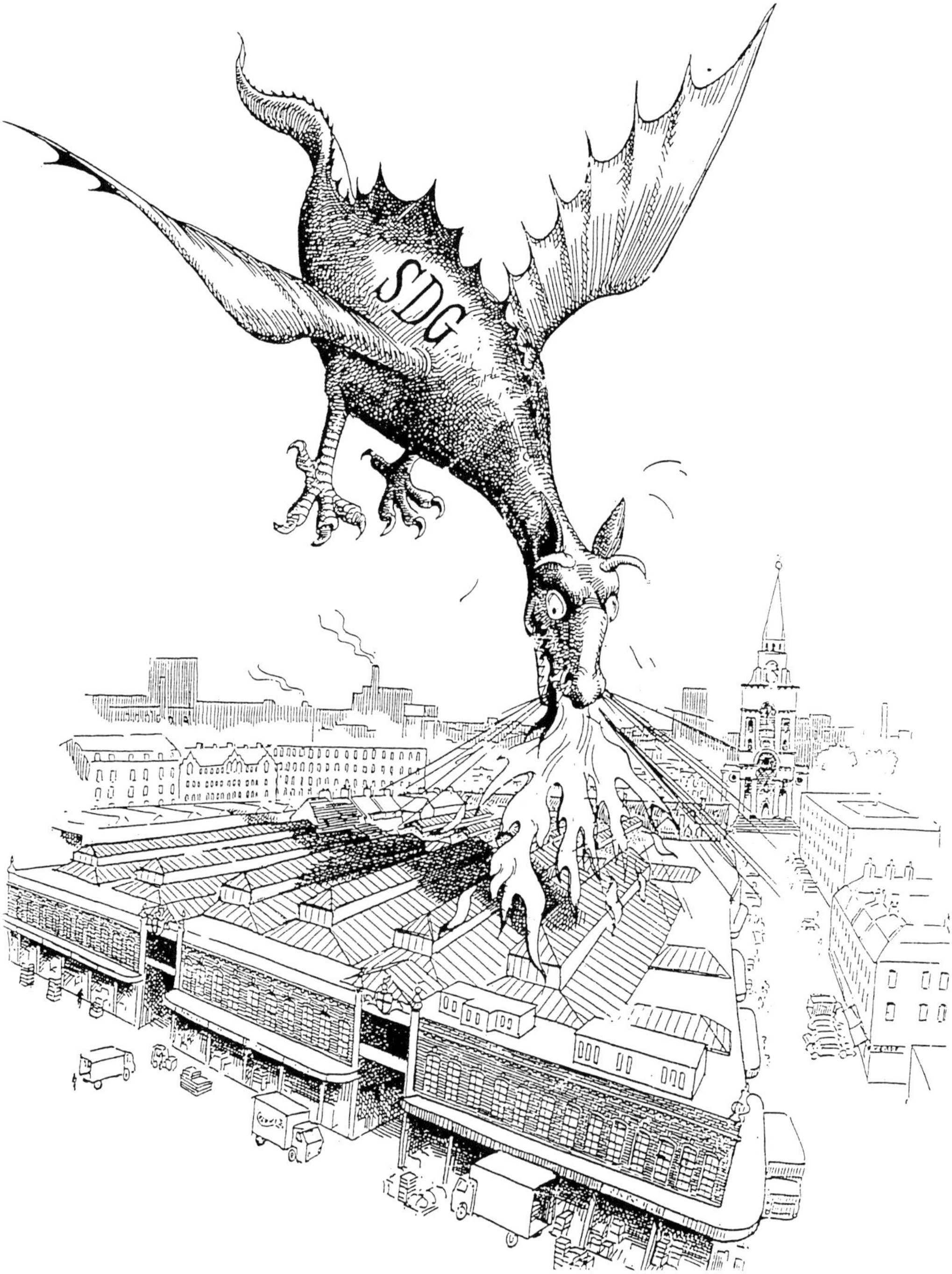
SDG

inquiries are more often than not arenas in which the design details of preordained development are negotiated.[4] In contrast, the CSSD's case against the Market redevelopment focused not on the architectural merits of the scheme but on the negative impact on the community of the relocation of the Market and the proposed development. Its concerns were with rising property values, preserving existing industries and jobs, and the broader issues of the City invading the traditional home of the underclass and 'revitalisation' being linked to free market development forces. Its arguments were unconcerned with issues of aesthetics and urban design, and it cared little for claims of the incompatibility of the Market and the residential enclave of the gentrifiers. The language and logic of the Campaign's opposition slipped outside of the discursive parameters of the public planning inquiry procedure.[5]

The local Left's campaign was built around a very specific notion of the character of the local community which it saw as threatened by the proposed redevelopment. Spitalfields was, in the eyes of the Left:

> a community of working class and industrious people: a multi-ethnic community ... a historic place which for over 3 centuries has harboured both refugees and immigrants.
> (Campaign briefing for Labour MPs, 1988, Records of CSSD)

Spitalfields, as depicted by the Campaign, was based on social cohesion, a family place where children played in the streets, women walked in safety and people were neighbourly (CSSD 1987 and 1988 Day 8: 47). The community mobilised by the Campaign appeared virtually untouched by the force of modern life. The Market itself not only stood as a signifier of a connection to a pre-industrial past but also as a buffer to the invasion of the City/modernity and the transformation of Spitalfields into 'one more line on the computer screen linking Wall Street and Tokyo' (CSSD 1987; see also Woodward 1993). Iris Marion Young argues that such models of community politics are posited as an alternative to 'impersonality, alienation, commodification' and rely on the premise that 'immediacy is better that mediation'. This is a fantastic and often nostalgic hope for, as I. M. Young (1990) notes, it is hard to imagine any aspect of modern life which might be 'unmediated'.

The presence of the Bengali population worked to authenticate the Left's evocation of the area as pre-capitalist, pre-modern and in need of protection from the City. In a book entitled *Spitalfields: A Battle for Land*, one former community worker and Campaign sympathiser said this of the area:

> With each migration Spitalfields has been charged by the struggle of

FIGURE 22
(*opposite*) The Campaign to Save Spitalfields from the Developer produced this cartoon of their view of the Market redevelopment. Here a griffin (the Corporation of London's fantasy zoomorph) emblazoned with the developer's initials has taken flight from the City and breathes fires of destruction over the Market site. For the Labour-linked Campaigners, the redevelopment of the Market site was read quite simply as the City (Big Capital) invading the home of the working class. (Source: CSSD Broadsheet)

village people in the vast metropolis – coming from small communities to one of the largest masses of humanity on earth. The village was self-sufficient. Spitalfields has been expected to provide the same self-sufficiency. Home, work, food, clothing, friends, relatives, doctors, schools, places of worship, markets must all be within walking distance which was the pattern back in the village. The demands of the village being stitched into the complex design of metropolitan life make Spitalfields a place of unique richness and variety.

(Forman 1989: 5)

In this description of Spitalfields, a local Bengali-specific politics of race is replaced by a narrative of immigration in which the pre-modern (village) confronts the modern (city). For the Left, the Bengali settlement enacted a 'natural' pattern for the area. From the earliest days, this was the home of marginalised ethnic minorities. Here Calvinist churches became Jewish synagogues and then Islamic mosques. The Huguenot weaving industry transformed into a Jewish fur and garment trade and then into the Bengali garment industry. Kosher butchers became halal butchers and Ramadan replaced Passover (Samuel 1989: 148).

The local Left's view of the Bengali community transformed a 'foreign' group into the 'natural' inhabitants of Spitalfields. Bengali residents were absorbed into an all-embracing 'indigenous' narrative of place which re-invented Spitalfields as a historic 'receptor for immigrants', a pre-modern place, which needed to be protected from modernity (CSSD 1989 Day 8: 25). The Campaign acted then as paternal protectors, not of the Bengali community *per se*, but of Spitalfields itself. Bengali residents of Spitalfields are both incorporated in and displaced by this paternalism. The Left is reinstated as the proper guardian of the inner city – not a working-class inner city but a multicultural inner city. But this new Spitalfields of difference often took forms that unsettled the 'pre-modern', anti-urban, communal nostalgias that gave affective drive to the Left's alliance with the Bengali community.

RE-INVENTING HOME

In 1980, when President Ziaur Rahman of Bangladesh visited Spitalfields, he was presented with a copy of a street sign for Brick Lane. He promised that on his return home he would rename an important street in Dhaka 'Brick Lane'. This is not an imperial naming of colonial territories, but a more obscure postcolonial return. 'Brick Lane' is taken back to Dhaka not because it signifies the British imperial heart, but because it signifies a thriving Bengali diaspora in that very heart.

For the most part the conditions in which the Bengali people dwell in Spitalfields are far from postcolonial. This is perhaps most clearly marked by the issue of housing itself. In Spitalfields 15.5 per cent of households are overcrowded compared with 2.1 per cent for Inner London and 1.3 per cent for Greater London. Over 75 per cent of Bengali households have been designated as overcrowded. This 'problem' of overcrowding is actually a problem of the mismatch between existing housing stock and the size of Bengali households: an issue that has generated both racist discourse about the nature of Bengali families and an anxiety about the reproductive capacities of this minority. Housing 'crisis' in the area is also registered in terms of the level of homelessness. In 1986, for example, more than 1,000 families were registered as homeless in Tower Hamlets. This problem was not the outcome of the racism of private landlords, but due to the borough's policy of identifying new arrivals from Bangladesh as having made themselves voluntarily homeless and therefore ineligible for government housing (Forman 1989: 231). The Bengali diaspora settles uneasily into the 'receptor' called Spitalfields.

The local Left's campaign against large-scale City expansion into the area spoke for an under-serviced and economically marginalised Bengali community, much like its traditional class constituency. But the Bengali community often slipped outside of the Left's constructs of it. There is, for example, a notable measure of Bengali business entrepreneurialism and some powerful sections of the Bengali community are active participants in the new-style, market-linked, community development initiatives which became a growing feature of Spitalfields under Thatcherism. This predominantly male, small business sector is politically powerful and the local Bengali institutions it has established and the community development alliances it has formed, directly challenged the Left's traditional hold over the political destiny of Spitalfields.

The distance between the local Left and these powerful sections of the Bengali community became patently clear in 1989 when Grand Metropolitan announced the closure of the Truman's Brewery operations in West Spitalfields and its intention to redevelop this site and the adjoining disused Bishopsgate Goodsyard. The 27-acre site straddles Brick Lane. A development scheme was proposed by the consortium of Grand Metropolitan, London and Edinburgh Trust and the British Rail Property Board. The aim was to create 'an urban village' which was not an extension of the City but 'separate from the City and able to thrive in its own right' through a mix of uses. Commercial elements, including large-scale office development were planned for the City side of the site. The remainder of the development combined housing, business space, workshops, industrial units and community facilities (London Edinburgh Trust and Grand Metropolitan nd: 1).

Perhaps with the lesson of the nearby Market redevelopment in mind, the developers of this scheme, from the outset, adopted a strategy of 'dialogue with the local community'. They established a Community Development Trust which was a partnership between the land owners, the community and the local planning authorities (see Fainstein 1994: 151–153). This Trust was to be directly responsible for any of the community planning gains generated by the scheme and certain sections of the site were to be passed into the direct control of the Trust. The developers were aware that their approach broke new ground. An LET/Grand Metropolitan spokesman called it a 'pioneering partnership' between developers and the community (*Guardian* 5 June 1990).

The Truman's Brewery/Bishopsgate redevelopment scheme evolved through consultation with a local organisation called the Spitalfields Community Development Group (CDG). The CDG was a local Bengali think-tank established by the male business sector in order to influence, but not halt, proposed development in Spitalfields. The CDG traced its origins to two new-style, market-linked, inner-city community development initiatives operating in Spitalfields. The first of these was the Spitalfields Task Force which was established as part of Thatcher's inner-city improvement programme and operated to invigorate local business and employment opportunities. The second was the Prince of Wales's Business in Community programme (BiC) which encouraged partnership between big business and small business in deprived areas. Both these new-style inner-city improvement organisations had a taint of a paternal imperialism, not surprisingly most clearly marked in the Prince of Wales's BiC scheme which, in his own words, was designed to bring some of the 'genius and skill of the City to . . . the needs of those who live virtually on the City's doorstep' (*Hackney Gazette* 3 July 1987). These groups, and the Community Development Group they gave rise to, were modelled on private/public partnerships typical of the Conservative Government's approach to inner-city revitalisation (Montgomery and Thornley 1990).

Using a £30,000 grant from BiC and with the encouragement of the developers, the CDG employed a team of planners to produce a 'community' scheme for the site. The community plan was based on local needs identified through an extensive series of community consultations and a 'planning for real exercise' in which local people discussed plans and models of the development. The plan emphasised appropriately designed social housing and affordable workshop space. The CDG saw the proposed redevelopment as an 'opportunity' rather than a threat and sought to achieve the 'maximum degree of community influence and involvement in the redevelopment process' (Community Development Group 1989: 4). The community scheme met with the approval of its indirect patron, the Prince of Wales.

The control of a portion of the redevelopment land, through a Community Development Trust, was a key plank in the CDG's vision of the development partnership (ibid.: 57). The CDG argued that for the local community to have any real power over and benefit from new development it must go beyond the usual planning gain arrangements of monetary contributions or housing and service allocations to the local community. The CDG requirement for control of at least part of the development land was an attempt to ameliorate the impact of the increased land values that would necessarily accompany redevelopment. The developers agreed to make over 11.5 acres of land to the east of Brick Lane to the CDG for social housing, workshop and retail units and community facilities.

The second distinctive aspect of the CDG plans included the development of Brick Lane as a 'Banglatown'; 'a vital and exciting focus of commercial and cultural life ... a bazaar area representative of the full range of Bengali, English, Jewish, Somali and other ethnic ingredients of the area'. Using existing buildings and appropriately designed new buildings the Banglatown would include shops, restaurants, showrooms and craft shops. It was hoped that this unique development would attract tourists to the area, and provisions were envisaged for bus parking and tourist interpretation points. Housing was included in the mix of uses which would constitute the new Banglatown for, it was argued, '[t]he presence of family and children will enhance the vitality and the authenticity of the whole area' (ibid.: 50). Banglatown was planned as a creative reinvention of another place, 'a recognition in the new of the values of history, but no slavish imitation of the past' (ibid.: 57).

Anderson (1991) has argued, using the example of 'Chinatowns', that cultural hegemony in multiracial societies is in part ensured through the sociohistorical construction of racialised categories in and through place. Conceptions of a 'Chinese race' became inscribed in institutional practice and pinned onto a locality known, and produced, as 'Chinatown'. At times this process was based on negative stereotyping of both the 'Chinese' and places called 'Chinatown'. More recently, and particularly within nation-states committed to multiculturalism, the process of making the social category of 'Chinese' through place continues through more positive, but no less essentialist, stereotypes. Anderson builds her argument through a social constructionist perspective which privileges external definitions of racial categories of Otherness, but acknowledges that this is never a 'simple process of cultural imposition on an unreflective audience' (K. Anderson 1990: 151). For example, in her consideration of recent efforts to consolidate and enhance Sydney's Chinatown, she notes that local Chinese merchants actively encouraged such development. In the emergence of the idea for a 'Banglatown' there was a similar process of local Bengali businessmen engaging in a bold and strategic mobilisation of essentialised, commercially viable and adequately consumable notions of being 'Bengali'.

The idea for a Banglatown was far from a simple 'return' or recovery of an ancestral past. Neither was it simply an 'appropriation' by external interests of essentialised notions of 'Bengali' in the service of economic diversification and expansion. Nor was it a clearly marked 'parodic subversion' in which essentialised identity categories were mimicked for the purposes of resistance (Butler 1992: 112). It was, rather, an activation of an essentialised identity category by one sector of the Bengali community within the terms of the enterprise-linked development opportunities available. The local businessmen who promoted 'Banglatown' traded on an essentialised notion of their culture as a component part of a broader plan to control redevelopment in their favour: acquire land, ensure social amenity and establish opportunities for Bengali youths to enter the 'City' workforce. According to a CDG spokesman, 'Banglatown' was an idea that helped to sell a broader argument for microeconomic development (interview, CDG member, 18 May 1994). It attracted attention to local economic aspirations because it packaged them in a racialised construct tuned to multicultural consumerism. The structures of identity marketed (quite literally) through the notion of 'Banglatown' worked to renaturalise and reconsolidate hegemonic notions of being 'Bengali'. They formed the cultural framework around which alliances could be made between big business and Bengali small business. There were subversive strands in this partnership, not least the claim for control of land and community gains from development. But by and large the essentialised identity encapsulated in 'Banglatown' provided the category through which bargains could be struck which would help to forge an economic home in a new nation.

The CDG had a patina of 'community' participation: it was Bengali in membership, it encouraged participatory planning, it advocated a form of urban land rights previously unseen in the area. Yet its strategy of forming alliances with big business and its idea for a Banglatown did not receive unanimous support. A Bengali Labour Councillor called the CDG plans 'naive and opportunistic' and claimed this group of businesses was far from representative of the entire range of community interests (Councillor Abbas Uddin, quoted in the *Guardian* 5 June 1990).[6] The local Left, which had uncompromisingly opposed the Spitalfields Market redevelopment, was particularly sceptical about the representativeness of the CDG. For the Left the CDG was not the 'real' Bengali community, but simply an unusually entrepreneurial, and almost exclusively male, section of that community.

The local Left found it difficult to reconcile the entrepreneurial bent of this powerful section of the Bengali community with its own imagined

Spitalfields of dependent minority groups. It was through this controversy that the limits of the Left's notion of Spitalfields-as-multicultural-receptor began to materialise. The Bengali businessmen were accused of trying to turn Spitalfields into 'another part of Bangladesh', a 'return' which was incommensurate with the 'tradition and history of Spitalfields' (interview with campaigner, 17 March 1989). The Left, as much as other interests, had an investment in confining Bengali identity to forms consistent with their own notions of Spitalfields as the emblematic site of an Englishness that accommodated, but then sought to domesticate, difference.

By the early 1990s the impetus for office development in Spitalfields and other parts of Inner London had receded. Neither the Market nor the Brewery sites had been redeveloped. The Market operations had relocated but the vacant site was placed into interim use which included a temporary sports facility (including a moveable swimming pool), a food hall (including organic fruit and vegetable stalls) and a mini-fairground. Developers awaited an economically more sympathetic moment to reactivate their grand visions. Here, as in the City of London, a complex politics of identity and place had been activated by buildings and spaces that never materialised.

In this politics of identity and place, various 'settlers' negotiated their own and others' rights to dwell. This was never a shrill exclusionary politics, although it did produce a dance of territorialisation and reterritorialisation. In Spitalfields the idea of community circulated as an imaginative category which attempted to gather difference together. Each of the groups claimed to 'know' the nature of the Spitalfields community, yet almost all celebrated its diversity and, by implication, the impossibility of such transparency (I. M. Young 1990). What became the 'knowable' trace for a Rousseauesque construction of a transparent community was not the Bengali population – nor even the Jews or the Irish or the Huguenots preceding them. Rather, 'knowability', a recognition of community, came by way of the historical idea of 'Spitalfields-as-multicultural-receptor'. In this imaginative construction, Spitalfields is not dislodged from the nation by the arrival of a postcolonial diaspora. Instead, this construct works to domesticate (not assimilate) the Bengali settlers within an embracing Englishness. This should not be thought of as a necessarily tolerant or benign encounter. Various versions of 'Englishness' are amplified within this 'multicultural' regime and often quite specific forms of being Bengali are required. The 'imagined community' of Spitalfields still produces its own spaces of exclusion and its own processes of displacement.

James Clifford (1994: 319) suggests that 'diaspora' is a predicament of the historicised condition of multiple locations. As Britain moves from empire to postimperial nation it is not just diasporic settlers who struggle to make a home. The non-Bengali rebuilding of a home-space in 'multicultural', postimperial Britain anxiously negotiates a nation full of Otherness. The making of homes and businesses in this nation is a process that anxiously attempts to create spaces within which the proximate Other might be ordered, sometimes harnessed, at other times domesticated. These habitually repetitious internal imperialisms establish the conditions within which Bengali settlers find ways of dwelling, of being Bengali, in the Heart of Empire.

NOTES

1   The socialist underpinnings of conservation faded from the agenda of later organisations like the Georgian Group (est. 1937) and the Victorian Society (est. 1958), which became increasingly elite in membership and more narrowly focused on the architectural merit of buildings. Even though 'new conservationist' groups like SAVE Britain's Heritage have, since the 1970s, expanded the range of buildings considered worthy of conservation and linked conservation with 'community' revitalisation, the class origins of the desire to conserve the built environment remain.

2   SDG is a consortium which included London and Edinburgh Trust PLC, Balfour Beatty Ltd (a subsidiary of BICC PLC) and Country and Districts Properties Ltd (a subsidiary of Costain Group).

3   The 118 social housing units were initially planned as on-site housing. More recently it was decided this housing should be located off-site. This change was justified on the basis of the delays in development, but the effect is to perpetuate a pattern of spatially managed cohabitation.

4   This was as evident in the struggle over Bank Junction as it was in the way the conservationists in Spitalfields, when they finally chose to, could convince state officials that the Market redevelopment was aesthetically inappropriate and needed to be subject to public planning adjudication.

5   Denied this forum, the Campaign took its formal complaints to two Parliamentary Hearings which were associated with the passage of a Bill to authorise the relocation of Market operations from Spitalfields to a new site.

6   The representativeness of the CDG became the source of considerable complaint. In 1990 an election was held to establish the management committee of the Community Development Trust which was to oversee the community development gains negotiated by the CDG. Before going to open election, the CDG struck a large proportion of nominees from the electoral roll. The Electoral Reform Society then declared the election null and void (Mythen 1991: 140).

5

URBAN DREAMINGS[1]

THE ABORIGINAL SACRED IN THE CITY

•

Before I shall become quite man again, I shall probably exist as a park.
(Miller 1966: 111)

Peter Weir's film *The Last Wave* (1977) opens with an Aboriginal elder painting sacred symbols on a rock face in outback Australia. For non-Aboriginal Australians this is a familiar image of Aboriginality: a male embodiment, a remote setting, a 'tribal' practice. The horror of *The Last Wave* comes from the uncanny movement of this spiritualised and remote Aboriginality into the modern city. A young Aboriginal man has taken some objects from a sacred site hidden under the architecture of Sydney. He is punished by his tribe, the victim of a 'tribal' killing, and the city is beset by the strange weather of an unleashed Aboriginal spirituality: an apocalyptic deluge of rain, hail, black rain and, finally, the 'last wave'. In *The Last Wave*, past time and distant people inhabit a modern Australian city. The film unsettles the familiar Manichaean geography of colonialism: the divisions of 'Western and ... Aboriginal, the modern and primeval, city and bush, settled and frontier ... the "known" self and the "unknown" other' (McLean 1993: 18).

On a bend on the Swan River, within view of the central business district of Perth, capital city of Western Australia, lies the Old Swan Brewery (Figure 23). Standing alone, flanked by the bushland of an urban park, it is a 'focal point' visible from the nearby city centre, a landmark for Perth residents as they speed past in their cars on their way from suburb to city. The distinctive group of nineteenth- and early twentieth-century industrial buildings is considered to be part of the European heritage of Perth. By the mid-1980s brewing had ceased on the site and the State government of Western Australia, through its newly formed Western Australia Development Corporation, planned to adapt the buildings for use as a tourist/leisure centre featuring restaurants, retail outlets and galleries. This vision was confounded by Aboriginal claims that the site was the home of the Waugal serpent.[2] From January to October 1989 Aborigines from the Perth region occupied this site

FIGURE 23 The Old Swan Brewery is located between the Swan River and Kings Park. A major road passes by the Brewery leading into the nearby city centre. Standing as it does surrounded by 'Nature' it is a prominent landmark for Perth residents. The Western Australia Development Corporation offered the development tender to a local firm called Brewtech. Peter Briggs, a 50 per cent shareholder in Brewtech, and one of the government's private business advisers, was later jailed for tax fraud

in protest against the proposed redevelopment. They appealed to State and federal agencies concerned with the protection of Aboriginal sites of significance and launched several legal appeals to the courts. The 'Aboriginal Fringe Dwellers of the Swan Valley', as they called themselves, not only opposed redevelopment of the Brewery buildings, they wanted the buildings pulled down and the area (re)turned to parkland. In 1980s Perth, the fantastic narrative of Weir's *The Last Wave* became strangely real. Residents and urban authorities in Perth were confronted with the unlikely presence of the Aboriginal sacred in the city.

The conflict that developed around the proposed redevelopment of the Old Swan Brewery takes my general argument about the (post)colonial politics of urban space to a site on the geographical edge of the former British empire. The contest over this place elaborates my case for understanding cities in terms of the ongoing structures and cultures of colonialism and the oppositional negotiations of its persistent force. The urban transformations that gave rise to the Old Swan Brewery conflict are most certainly a product of a property and development boom and a diversification of capital accumulation into industries of consumption. In this sense, the contemporary trajectory of this site is connected to global economic transformations which, in part, transcend the spatial, political and economic logic of nineteenth-century colonialism. Yet this most 'global' development negotiated a very specific local politics deeply marked by the historical legacy of the colonial dispossession of indigenous peoples.[3]

ORDERING THE URBAN

The colonisation of Australia was based on establishing a white settler colony in a land previously occupied by an indigenous peoples, the Aborigines. The desire to establish settler colonies depended upon the will of erasure or, when this failed, systematic containment of indigenous peoples. In the case of Australia, this 'erasure' was inaugurated by the notion of *terra nullius*, land unoccupied, which became the foundational fantasy of the Australian colonies. The justice of *terra nullius* was debated in the British Parliament and its truth was daily challenged by the undeniable presence of Aborigines in the colonies. From the outset it has been a most unstable foundation for the nation. Its tenuous and debated reality was necessarily shored up by a whole range of spatial technologies of power such as the laws of private property, the practices of surveying, naming and mapping and the procedures of urban and regional planning. In the areas where cities grew, those areas of first settlement, quick and comprehensive colonisation was necessary to create the material preconditions for the realisation of a 'settler' colony. Towns provided the home for settler 'authority'; they were administrative hubs of colonial governmentality and orchestrated the acquisition and redistribution of material resources. From these secure administrative centres settler expansion could move outward into those lands where the fantasy of *terra nullius* was less surely inscribed.

The 'city' of Perth was ritualistically marked on 12 August 1829, the birthday of King George IV, when the Lieutenant-Governor of the colony, James Stirling, led a small group to the elevated ground chosen for the city. A Mrs Dance was given the honour of striking the first blow to a tree which was felled to mark the commencement of the comprehensive land clearances which made room for the city to be called 'Perth' after the Scottish city of the same name (Green 1984: 51). In this event of colonial commencement, a woman took the symbolic edge of/off the masculinist conquest of territory. Like other cities in the emergent colonies of Australia, Perth was based on British colonial planning wisdom. An ordered grid was emphatically placed over the land and provided the spatial skeleton for an embodied settlement (Figure 24). Colonisation brought the city to 'a country without cities' and constructed 'little copies' that mimicked the imperial heart (Muecke 1992: 5). Paul Carter's spatial history of Australia has shown how the logic of the grid, most clearly and immediately articulated in relation to the townships of colonisation, worked like the map to connect and give unity to space 'in advance' (Carter 1987: 204). Gridded urban plans were undoubtedly imported spatial orderings intended to realise colonial authority. Yet, as Carter argues, their symbolic functioning was far from stable. The grid did connect Australian cities to a wider order of power which led back to the imperial heart

CITY OF PERTH
PERTH
PERTH STATION
WELLINGTON ST.
HAY ST.
ST GEORGES TER.
MALCOLM ST.
MOUNT ST.
KINGS PARK ROAD
OBSERVATORY
MT ELIZA
Pt. Lewis
THE NARROWS
Pt. Belches
PERTH WATER
SOUTH PERTH
No 127
No 115
RIVERSIDE
GOVERNMENT HOUSE
TOWN HALL
ADELAIDE TER.
Lot. 563
Park & Recn
(Gudinup)
(Gabbykelga)
(Kooyamulyup)
(Gareenup)
(Buneenboro)
(Mata garup)
(Gabudjulup)
(Goorgoogoo)
(Karragatup)
Reservoir
Jetty
Beacon
Quarry Pt. Beacon
Market Place
DELHI SQ.
MOORE ST.
HAVELOCK ST.
ABERDEEN ST.
JAMES ST.
ROE ST.
WILLIAM ST.
BEAUFORT ST.
STIRLING ST.
SHORT ST.
ROYAL ST.
WICKHAM ST.
BRONTE ST.
QUEENS GARDENS
EAST PERTH STN
Rainfall 34"
Var. 4°20'W 1905

but the local purchase of this authority was a more ambiguous matter. Colonists 'succumbed' to the gridded city, but not because they were responding to the orders of imperial authority. They settled comfortably into these cities because the grid was recognisable, it was a familiar spatiality in an unfamiliar land. It was for them 'the traditional matrix of new urban beginnings' (ibid.: 210). For Aborigines, the corners and lines which began to be carved into their land did not herald a familiar order that need only be awaited. This new geometry quite literally marked an unknowable future of imperfect encounters with those who sought, ceaselessly, to realise the perfection of the grid.

Aborigines were both necessary and problematic in the colonial occupation of Australia. 'Exploration' of the Australian continent was most often under the guidance of Aborigines and settlement often traced the spatiality of Aboriginal knowledge of available water and pasture. But 'possession' of the Australian continent required the Aboriginal presence, at the very least, to be contained and, at its most thorough, to be eradicated. From the moment of 'settlement', the Australian nation has realised itself through regularly intractable and frequently violent interactions between Aborigines and colonial settlers. The case of south-west Australia was no exception. Between 1826 and 1852, there were 178 recorded violent encounters between Aborigines and colonisers. In these conflicts 108 Aborigines and twenty-seven colonists were killed and forty-four Aborigines and thirty-two colonists injured (Green 1984: 203–218). Most of these conflicts emerged from Aboriginal thefts of food, for colonisation robbed Aborigines of access to traditional food sources. Other conflicts were the outcome of colonists interfering with 'fearful' Aboriginal ceremonial activities. The early colonial ordering of urban space, the making of Perth the city, sought to eliminate indigenous people and indigenous meanings, those 'local authorities' which might compromise the mimetic transfer of imperial urban models to the colonies (de Certeau 1984: 106). Aborigines were positioned outside the 'civilisation' of the emergent city: a pure negativity against which settlers constituted their sense of Self. If not eradicated, then they were to be kept 'distant' by a process of spatial containment. Urban planning became the vehicle for what Bauman (1992: xv) describes as the 'perfect world that would know no misfits . . . no disorder . . . no vagabonds, vagrants or nomads', where there were 'no unattended sites left to chance'.

The land upon which the Old Swan Brewery was later built played an important role in the initial measures used in Perth to contain Aborigines. It was at this site, in 1833, that the colonial government established the Mount Eliza Depot for Aborigines (Vinnicombe 1989: 22). 'Public' access to this space was prohibited and it was hoped that Aborigines would settle

FIGURE 24
(*opposite*) Part of map dated 1921 prepared by the Department of Land Administration, Western Australia. In this early map of Perth the emphatic placement of the straight lines and corners of the grid city is clearly shown. On this map the indigenous dwellers of the land that became Perth are but a parenthetic trace of Aboriginal place names. 'Goonininup', the Aboriginal name for the Waugal Ground, is marked in the narrow part of the river just below Mount Eliza. (Reproduced by courtesy of the Department of Land Administration, Western Australia)

permanently at the depot. But the colonial will to contain Aborigines and keep them separate contradicted the material needs of settlement and underestimated the reluctance of Aborigines to 'enjoy' colonial protection. Segregation was never fully realised and quickly gave way to more disorderly and permeable spatial arrangements. The Governor of the colony expressed his displeasure at the inability of the new Depot to keep 'the natives' off the streets:

> the natives of the Perth Tribe have become of late extremely troublesome and even offensive to the inhabitants of the town by going about quite naked and by tumultuous and hostile meetings in the streets.
> (*Perth Gazette* 22 February 1834, quoted in de Burgh and de Burgh 1981: 124)

By the early twentieth century, Aboriginal prior occupation was only vaguely traced on the official maps of Perth. The Aboriginal name for the Old Swan Brewery site, 'Goonininup', survived on a 1909 plan of the city but only in parenthesis (Vinnicombe 1989: 16 and appendix 2; Mickler 1991a: 6; see Figure 24). The parenthetic gesture of the map aptly signifies the way in which Aborigines had come to inhabit Perth. The provisions of the Native Administration Act 1905 (WA) ensured Aborigines were officially banned from the Perth Metropolitan Area until 1954 (Churches 1992: 10). An Aboriginal presence was judged to disrupt the visual and social ordering of the city. One of the Aborigines protesting the redevelopment of the Brewery site recalls how this regulation was enforced:

> You wasn't allowed to camp around there at Kings Park by the river on our Homegrounds. Police would shift you even if you sat on their seats even round at the Esplanade. They only wanted bridiahs (upper class white people) there. You wasn't allowed to lay around there at Kings Park. The old people used to talk about Kings Park, all along the river, but you wasn't allowed to stop there.
> (Reggie Wallam in Ancestors of the Swan River People 1989)

Aborigines were not eliminated from the Perth scene, but lived on designated reserves, in state-allocated houses or as 'fringedwellers' in informal camps in 'unseen' parts of the city. By the mid-1950s the early segregation and protection policies towards Aborigines gave way to an official policy of assimilation. Assimilation was intent on orchestrating the 'disappearance' of Aborigines through their absorption or integration into 'mainstream' Australia. In Perth, as elsewhere, Aborigines who were considered 'white' enough (a judgement based largely on skin colour) or 'civilised' enough (a judgement based on their adoption of European ways) were encouraged to leave reserves and move into suburban housing modelled on the British notion of the Garden Suburb. State-owned houses, strategically scattered through the lower income areas of cities, were offered to Aboriginal families in the belief that spatial integration would lead to social integration. The conditions under which Aborigines could 'dwell' in the city of Perth forced them, by exclusion or assimilation, to become an 'invisible', but menacingly present, citizenry.

VISIONING DEVELOPMENT

The boom times of the 1980s were not confined to global cities like London. Throughout the 'western' world more modest cities experienced the transforming force of industrial restructuring, the deregulation of financial sectors and property speculation. The city of Perth, with a population of under one million and the most remote of the urban centres of Australia, was one such city. During this period a new breed of businessmen emerged in Perth. Among their ranks were men like Alan Bond who, after arriving from Britain on a £10 assisted passage to become a 'New Australian', managed to make good. Like Alan Bond, many of the new breed of businessmen were global players. Their global games of speculative investment and conspicuous consumption were traced onto the local terrain of Perth through a range of entrepreneurial ventures.

The entrepreneurs of Perth had a sympathetic partner in the new Labor Government which sought to encourage partnerships between the state and the distinctively buoyant and personality-ridden private sector. The Labor Government's vision was to put Perth 'on the world map' through the promotion of the 'sunrise' industries of tourism, finance and high technology, although to a large degree the wealth of the State remained locked into mineral and resource exploitation. Alan Bond's 1983 victory in the America's Cup yacht race, and Perth's subsequent role as the host of this global sporting event, gave considerable impetus to the shift to culture- and leisure-based industries. The period was marked by a radical restructuring of the boundaries between the private and public spheres and an active promotion of cultural development through private capital investment. Under the new-style Labor Government, local business entrepreneurs became formal government advisers and private/public entrepreneurial agencies, such as the Western Australia Development Corporation, were formed and allowed to operate outside of established parliamentary and planning procedures (Alexander 1992).[4]

The redevelopment of the Old Swan Brewery was a small but controversial component in this larger process of restructuring. From the moment the site came onto the market in 1978, private development aspirations and notions of public amenity were blurred. The site was originally purchased by a private developer for A$4 million. But after public concerns were voiced about 'insensitive' foreshore development, the government purchased the land for 'public use' and declared it a Reserve for the purpose of government requirements. Under this designation the land was removed from the jurisdiction of normal town planning regulations. Within the entrepreneurial logic of the moment, 'public amenity' was readily translated into a government-led tourism and service industry development. In 1986 the Western Australian Development Corporation announced the site would be developed by the brewery and hotel chain, Brewtech.[5] Initial plans included office space, a 450-vehicle car park, a 500-seat theatre, a museum display, a boutique brewery, various 'multicultural' food outlets and a 'genuine Aussie pub'.

URBAN NOMADISM

The Aboriginal protest against the Old Swan Brewery redevelopment was led by Robert Bropho. He describes himself as a 'fringedweller' and has spent most of his life living on the 'margins' of the city (Bropho 1980). His precarious dwelling in the city was shaped by the historical fact of dispossession and his own negotiation of the permissible places for Aborigines to be, namely on institutionalised reserves, in allocated housing or homeless. For most of his life he lived in places inbetween; vacant land on the urban fringe, lands which were the marginal spaces, the 'dirty' spaces, of the city.

> I've been a fringedweller all my life in and around the metropolitan Perth area, living near the local junk tip at Eden Hill, under sheets of tin, lived with my mother and father in the Swan Valley area in places such as Widgie Road, South Gilford reserve before it became Allawah Grove, lived in a camp at Caversham in the mid-thirties, lived in bush breaks at Bayswater, tin camps at Swanbourne, lived on the fringes of Allawah Grove when ... the place [was] policed [by] a white adviser to a black cause.
>
> (ibid.: 1)

As a youngster with his family, fringedwelling was not by choice: Aborigines were the 'banned' people of the city. Under assimilationist policies which encouraged Aborigines to live by Anglo-Australian protocols of urban dwelling, Bropho's fringedwelling became a strategic refusal of state regulation. It provided a way for him and his family to 'return to where we knew we belonged ... back into the past where we came from' (ibid.: 56). Through fringedwelling Bropho removed his family from the expectations of assimilationist suburban living and the everyday prejudices of life in predominantly Anglo-Australian suburbs. Bropho's fringedwelling was a way of avoiding the suburban version of what hooks (1992: 344) has described as the 'terrorism of whiteness'. For Bropho the closure of Aboriginal reserves and the promotion of assimilation was just another phase of colonialism which took from Aborigines the little land they had left, the final gesture in a process which had rendered him and his people 'a race ... without a country' (Bropho 1980: 15).

In Australia, efforts to acknowledge Aboriginal land interests and to provide some form of rights over land began in the 1970s. These provisions are notoriously inadequate in terms of Aborigines with interests in urban lands. The majority of lands returned to Aboriginal ownership have been in central and northern Australia, far from the urbanised coastal regions. This specific geography is a product of the limitations within various land rights provisions. On the one hand, the lands designated as available for claim are largely those not currently leased for productive use such as existing Aboriginal reserve lands, vacant Crown lands or national parks. Urban land, which is incorporated into multiple and productive uses and is largely under private ownership, is rarely available for land claim. On the other hand, the more comprehensive and potentially generous of the land rights provisions have been biased in favour of those Aboriginal groups who can prove or have an undisputed traditional way of life and association with the land. Land rights success in Australia was, and still is, linked to specific constructions of Aboriginality in which 'traditional' Aborigines are privileged over those Aborigines who have had their way of life most seriously disrupted by contact (Jacobs 1988).

For urban Aborigines rights over land have been piecemeal and incomplete. Their interests in land have been recognised, in part, through state-sponsored land purchase programmes working within the limits and opportunities of the private land market. Alternatively, urban Aborigines have had land interests recognised through programmes that record and give legal protection, but no proprietorial rights, to sites of cultural significance. Even within the provisions of site recording programmes, it has been those Aboriginal communities who conform to traditionalist notions of Aboriginal associations with the land who have been best served. The emphasis of much site recording and protection policy was initially on archaeological sites or sites that were significant in terms of clearly traditional practices and beliefs. Only recently have site registration and protection mechanisms expanded to include a more flexible notion of what a site of cultural significance might be for Aboriginal communities.

The limits of existing land rights provisions were unbound in 1992 when the High Court of Australia handed down a decision that found in favour of Eddie Mabo's claim that his people's entitlement to the Murray Islands in the Torres Strait had not been extinguished by the settlement of Australia by the British. In reaching this decision the Court held that the common law in Australia recognises a form of native title which is also applicable to mainland Australia. This decision displaced the fallacy of *terra nullius*, land unoccupied. It opened the way for the implementation of the Native Title Act 1993 (Lavarch 1994: iv). The Mabo decision implied that all lands, including urban lands, were once legitimately native lands and potentially open to claim. It heralded the possibility of a more equitable and uniformly applied land rights provision, including the possibility of meaningful urban land rights.

The requirements of the Native Title Act suggest that in practice those Aboriginal communities which can prove a traditional and uninterrupted interest in land will still be best served by the legislation and that lands not under long-term non-Aboriginal use will remain difficult to claim. That is, it is unlikely that Native Title will open the way for claims over urban lands.[6] Despite these limitations the Mabo decision generated an intense anxiety among many non-Aboriginal Australians, especially around the possibility that – in principle if not in practice – claims might be made over city and suburban lands. As one press editorial noted:

> There is a significant gulf between recognising Aborigines' rights to something like Ayers Rock [*sic* – Uluru] and to allowing them open slather ... encouraging militant Aboriginal groups to lay claim to sites such as the heart of Collins Street in Melbourne or Bennelong Point in Sydney.
> (*The Australian* 23 July 1989: 3)

The Mabo decision and Native Title Act dismantled the established spatial architecture of existing land rights provisions in Australia which comfortably placed a spiritualised, 'tribalised', land-rights deserving, Aboriginality well away from the urban centres.

The disturbance of this moment comes not simply from the possible reterritorialisation of urban space by Aborigines, but also from the unpredictable and often unknowable sacredness such claims bring with them. Richard Sennett (1990: 16) has argued that the development of the modern western city saw the marking-off of sanctioned and ordered sites of the sacred from the threateningly disorganised spatiality of the secular.[7] Kong (1993), using the example of multi-religious Singapore, shows how the religious geography of the contemporary city is controlled by the secular state in much the same way as any other land use. In the contemporary city sacredness is planned for, not to protect it from the threat of the secular but because the secular, the accumulative drive of the urban, needs to be protected from the irreverence (unproductivity) of the sacred. Yet such divides are always fragile. As my reading of the struggle to redevelop Bank Junction in London showed, even in the most modern of cities the spatially discrete sites of the sacred are complexly intertwined with the secular imaginings of place and nation (see also Daniels 1993).

The Aboriginal sacred is deeply antagonistic to urban modernity's need to keep the sacred apart from the secular and to regulate it as if it were just another land use. This is in part because of its 'hidden' quality. Aboriginal sites are often subject to strict protocols of disclosure and many are 'secret' sites known only to properly authorised individuals. The 'hidden' status of the sacred is also, and in a paradoxical sense, elaborated by a history of non-Aboriginal repression – the failure of colonial Australia to recognise and give legal status to such sites. In recent years various provisions have been made to recognise and protect, to recover, Aboriginal sites. The official procedures require registers of sites to be created in state archives. That is, these procedures take previously 'unknown' (not officially recognised) Aboriginal land-based knowledge and place it into the public domain. Recording Aboriginal land-based knowledge, cultural or sacred sites as such knowledge has come to be known, establishes the preconditions for a power/knowledge nexus which rests not with Aborigines but with those state agencies that build 'complete' and spatially fixed reconstructions of this knowledge. In various ways, Aboriginal groups have attempted to negotiate a balance between the present pragmatic necessity of disclosure, the need to register sites so that they are afforded legal protection, and the necessity of secrecy, the need – both traditional and strategic – to keep their land-based knowledge to themselves.[8] Some communities have even engaged in a process of strategic non-disclosure, in which land-based knowledge is not disclosed unless a site is under direct threat. This strategy establishes a specific chronological and geographical articulation for land-based knowledge. Sites remain 'unknown' until under the threat of destruction. This structure of articulation itself generates a complex politics of authenticity and validation. Sites of significance to Aborigines can, quite literally, just 'appear' under the force of a development threat and are extremely vulnerable to charges that they are simply 'invented traditions'.

Both site-recording procedures and Aboriginal strategies of non-disclosure create conditions in which something once repressed or secreted suddenly manifests itself. Modern Australia is now haunted by the possibility that 'new' sites will be discovered or made known. In this sense, the Aboriginal sacred is omnipresent in its absence (Taussig 1987: 78). The Aboriginal sacred has a 'nomadic' geography which is not derived from a premodern character but is produced by the conditions of articulation established under modernity. The possibility of sacred space simply appearing, coming from 'below ground' as the 'hidden' Aboriginal sacred appears to do, is radically unsettling to contemporary practices of allocating space.

Such a haunting occurred in Perth during the 1980s when, in response to the new boom of urban redevelopment, it was decided to conduct the first systematic surveys of Aboriginal cultural, archaeological and sacred sites in the city. Two surveys (1984 and 1987) of ethnographic and archaeological Aboriginal sites in the Perth area were made by the Western Australian Museum, the official state agency with the responsibility of protecting Aboriginal cultural properties (Vinnicombe 1989: 27).[9] The 1987 Museum report specifically sought to insert 'Aboriginal sites' into the planning equation of Perth development:

> It would be highly desirable to save all Aboriginal sites.... This is an impractical idea however.... Assessments of significance attempts to determine the most important sites (to various interest groups), or the most representative or most informative, and save these.
>
> (Strawbridge 1987/8: 18)

From the outset, it was hoped that the recovery of the repressed Aboriginality of Perth could be contained by the limits imposed by a combination of planning pragmatism and 'representative' categorisation. Precise mapping of sites of significance was seen as a specific requirement for their compatible incorporation into the framework of planning for urban space. As the report stated, 'for the purposes of management some real, physical boundaries must be established' (ibid.: 21).

Mapping Aboriginal sites of significance within western cartographic logic is a notoriously troubled project (Jacobs 1993). Ken Maddock (1987: 135) argues that whatever the spatial extent and location of a site of significance, it cannot simply be protected by a perimeter fence. This inability to be constrained is in part a result of the nature of the sites themselves and the way in which their significance 'radiates out from them', making them more like 'smudges' on a map than 'pinpoints'. But, as I argue later, it may also be a product of the way in which these sites are articulated in and through modernity itself. Whatever, the desire to map Aboriginal sites precisely responded to the spatial logic of planning for development rather than an unfettered recovery of Aboriginal interests in land.

Despite its measured and pragmatic approach to the recovery of Aboriginal site-based interests, the Museum reports excavated a local knowledge that shook the foundations of modern Perth. The Museum sought to define and confine Aboriginal land interests in Perth but, in uncovering these sites and placing them into a public sphere of knowledge, it produced the opposite effect. For example, one of the very sites 'uncovered' was the Waugal Ground at the Old Swan Brewery site.[10] The Waugal's sudden 'reawakening', as the media preferred to call it, heralded an anxiety about an 'illegitimate proliferation' of sacredness (*The Age* 22 February 1989; see also Gelder 1993: 500). One newspaper reported that the claims about the sacred site of the Waugal Dreaming was just the 'myth at the tip of the iceberg of claims' and calculated that of an estimated 13,000 significant sites throughout Australia at least 1,100 were in the Perth metropolitan area (*The West Australian* 22 June 1989: 1).

THE EROTIC CITY

Nineteenth-century Perth was a mimetic reproduction of imperial London, but twentieth-century Perth is just as likely to be compared with 'global' cities like Hong Kong or New York. The core of contemporary Perth is self-consciously modern in its architectural form. The skyscrapers of the city stand as markers of the wealth of the State derived in large part from extractive industries. From a heritage viewpoint, Perth is 'just another late 20th century western-style city of skyscrapers' (Heritage Council of Western Australia 1991: section 4.8.2).

Government provisions to protect the historic built environment of the city came relatively late to Perth and by the early 1980s Western Australia was the only State in Australia without such legislation.[11] The reluctance of the Western Australian Government to introduce its own historic buildings legislation was closely tied to the pro-development mentality of the State. It was not until 1987 that the Western Australian Parliament attempted to make specific provisions for the registration and protection of non-Aboriginal heritage and not until 1990 that legislation was passed to create a Heritage Council with special responsibility for buildings and areas of special significance.[12]

Perth was not without its advocates for the protection of the historic built environment. The National Trust of Australia (WA) created its own list of buildings of historical significance but enjoyed only limited advisory powers within the planning system. Other city promoters, planners and architects made a case for local planning regulations to protect the historic built environment. One 1960s publication urged Perth planners to adopt British models of urban conservation and offered Sir William Holford's postwar plan for the rebuilding of the City of London, which combined a conservation sensibility with the science of town planning, as a suitable model. Lord Esher, then President of the Royal Institute of British Architects, gave his endorsement to this vision for the city on the edge of the former empire (Oldham and Oldham 1961: i and 91). This 1960s vision for a heritage future was at the same time a colonial return. In the 1980s a more explicitly Australian heritage vision for Perth was articulated by a group of city promoters and planners known as CityVision. Unlike their 1960s predecessors, the CityVision group did not seek to restore a colonial past but to celebrate a 'multicultural' present through creative partnerships between private capital and culture. Mickler (1991b: 77–78) argues that these multicultural planning visions are 'little more than an exotic post-colonial urban reverie' which produced regulated and sanitised 'sensual delights' of food and performance for Anglo-Australian consumption.

Throughout self-consciously multicultural Australia urban areas which bear the marks of colonial and migrant settlement are being eroticised and exoticised through planning intervention. Such interventions exemplify what is an increasingly common feature of contemporary urban planning – the recourse to urban design as a planning solution. Through such multicultural planning, a politics of difference – which is also the uneven politics of race – is aestheticised. The problematic of such processes is not whether such planning measures and developments commodify or aestheticise the 'real' but whether, as Iris Marion Young (1990) would argue, they work for or against a productive politics of difference. In the case of Perth, built environment heritage retrievals and multicultural celebrations produced particular parameters for cultural valorisation. Aboriginal aspirations for land rights in this more 'civic', but also more eroticised and aestheticised, Perth were defiantly outside of these parameters.

BREWERY DREAMINGS

The plans to redevelop the Old Swan Brewery buildings were consistent with global processes of revalorising heritage buildings and incorporating them into the new service industries of tourism and leisure. Indeed, the classification of these buildings as part of the non-Aboriginal 'heritage' of Perth was inextricably entwined with development aspirations for the site (Figure 25).

When development was initially proposed there had been no official recognition of the heritage status of the buildings on the site. In 1986, soon after the unveiling of development plans, the National Trust of Western Australia conducted an assessment of the site and declared the Brewery and surrounding area as a Historic Site and classified the old stable, which was to be demolished in the redevelopment plans, as a historic building in its own right (National Trust of Western Australia 1987). The Trust warned against over-development but encouraged the government to proceed with 'the presentation of the site and its surrounds as an historic site as well as a major landmark for the City and the river' (National Trust of Western Australia to Minister for Education and Planning 6 February 1986).[13]

Throughout the 1980s there was a systematic elaboration of the initial and tentative assessments of the buildings' heritage value. An alliance of concerned individuals, acting under the title of the Brewery Preservation Society, argued that the buildings were significant because of their outstanding architectural merit and also because of their value as 'working class heritage' (*The West Australian* 10 July 1989). It was through the lobbying of the Brewery Preservation Society that the buildings on the site were given legal heritage status in October 1989 by way of the federally administered Australian Heritage Act. When, in 1990, Western Australia finally passed its

FIGURE 25 As soon as the Western Australian Government had passed the necessary legislation to protect its settler heritage the Old Swan Brewery was formally listed. Shortly after, a sign was erected on the fence surrounding the still undeveloped Brewery. Although the Waugal ground had also been registered some years earlier under Aboriginal site protection legislation, the government erected no sign to mark that legal status

own heritage legislation, one of the first major assessments dealt with the Old Swan Brewery buildings. In this sense, the proposal to develop the Old Swan Brewery as a tourist and leisure site actually preceded, indeed helped to produce, the official heritage designation.

In official assessments of the social relevance of the Brewery, the buildings came to mark an 'Australian ethos' of 'beer, work, tourism and sport' (Heritage Council of Western Australia 1991: section 4.8.2). Western Australians, the official assessment noted, were once ranked among the biggest consumers of alcohol in the world and could proudly claim to have drunk twice as much as their nearest rivals in Queensland. Swan brand beer is, in the words of the report, 'synonymous with Western Australia'. Once this may have been a local assemblage of a beer brand and State identity, but now it is registered on a global scale. Under the guidance of Alan Bond, Perth's favoured man of 'beer, work, tourism and sport', Swan beer has long departed from its historic site of production on the banks of the Swan River and entered a global circuit of production and consumption. It is not simply the globalisation of the Swan brand that unsettles official claims about the 'Australianness' of this site. Beer itself is less a signifier of a unified 'Australian ethos' than it is of a masculinist and racist 'ethos' of being Australian. For Aborigines in particular beer has marked not the forging of identity, but the ravaging of identity. Alcohol dependency among sections of the Aboriginal community has provided the basis for the elaboration of negative stereotyping of Aborigines and is now considered by Aborigines to have contributed to the breakdown of their traditional culture. Alcohol also marks the contingent nature of Aboriginal citizenry in the Australian nation: Aborigines were not legally entitled to buy alcohol until 1967 when they were finally given full rights of citizenship.

The elaboration of the cultural value of the Old Swan Brewery also came by way of townscape assessments. Located as they were in a relatively isolated setting on the banks of the river and with an urban park as backdrop, the Old Swan Brewery buildings readily succumbed to visual revalorisations of their 'landmark' quality. The site was marked as a 'tranquil' space of the past, which acted as a counteractive gateway for the modern Perth of 'freeways and highrise buildings' (Gregson 1989: 37); a 'picturesque contrast' to and a 'dramatic framing' for the 'developed' areas of Perth city (Heritage Council of Western Australia 1991: section 4.5.2). A 1991 assessment of the site concluded that the buildings evoked a romantic European scene of 'ancient structures clinging to the side of a waterway' (ibid.). The Old Swan Brewery was imaginatively reinvented as the heritage marker for a city that historically had had little time for the past. The site was quickly incorporated into what Urry (1990) refers to as the 'tourist gaze'. Jolly Jumbuck and Captain Cook cruises passed the site, but their commentaries were just as likely to speak about a present sensationalised by Aboriginal protest as a past tranquillised by the tourist gaze.

Under the conditions of development the Old Swan Brewery buildings were produced as non-Aboriginal 'heritage', a place where cultural and economic value excited each other. This revalued space was set against Aboriginal claims about the sacredness of the site and their aspirations to see the land returned to 'nature'. Foucault (1986: 23) suggests that the inviolable spatial oppositions of contemporary society are those that are underpinned by the 'hidden presence of the sacred', that which is valued. This spatial controversy emerged not only because an incommensurate sacred and secular met, but because of a sudden proliferation of sacredness: the production of sanctioned heritage which was closely tied to development aspirations, alongside the sudden 'appearance' of an Aboriginal sacred which was irreverent to development. This was, as the then Premier of Western Australia suggested, an 'unfortunate clash between European heritage and Aboriginal heritage' (Western Australia House of Assembly 1990: 1659).

PLACING THE WAUGAL

The anxiety generated by the 'appearance' of the Waugal took him well out of his specific geography on the banks of the Swan River: it helped to produce the unbound geography of threat. Aboriginal claims about the Old Swan Brewery site possessed the city, invaded the very discursive constitution of the city:

> hitherto only vague murmurs, of Waugal tracks, of sacred water, of ceremonial sites, began to circulate in court-rooms, government offices, Council Houses, churches, union offices and in the media and thereafter in schools, supermarkets, taxi cabs and dinner parties. Across innumerable discursive environments ... [in a] ... veritable industry of sense-making.
>
> (Mickler 1991a: 2)

In contemporary Australia the idea of the Aboriginal sacred is amplified by the very measures used to regulate and constrain it. It is through these procedures that the Aboriginal sacred enters into the public domain and becomes what Ken Maddock (1988: 305) refers to as a part of the very (uncomfortable) 'furniture of the Australian mind'. The awakened and unleashed Waugal activated a frantic activity of (re)placement. As Mickler

(1991a: 2) notes, it produced a 'charting of space and time between the poles of two cultures'. The flurry of mapping 'anxiously repeated' the task of knowing and identifying Aboriginal interests in order to keep them 'in place' (Bhabha 1994: 66).

Placing the Waugal precisely in space proved difficult. The various commissioned reports on the significance of the area to Aborigines agreed that 'the entire area along the base of Mount Eliza ... was of considerable significance [to Aborigines] both economically and spiritually'. Waugal was known to be associated with a number of specific landscape features in the vicinity of the Old Brewery, such as a local spring and some large stones, but officials could not verify which water source and which stones. It was admitted that the area known as the 'Goonininup' camp ground was 'not capable of precise identification and possibly never was' although boundaries were still determined using archaeological and historical records (Senior 1989: 13). Such anxious remappings of Waugal were attempts to reassert the familiar order of colonial spatial logic in the face of the disorderly emergence of a repressed geography of Aboriginal meaning. It was an attempt to restore order through an incorporation of Aboriginality into a permitted and more systematic geography of the city.

The 'failure' of Aboriginal land interests to conform to the rigid spatiality required by cartographic translations has always had the capacity to serve the interests of those wishing to undermine Aboriginal land claims. In the case of the Old Swan Brewery claim, the project managers for the redevelopment scheme actively highlighted Aboriginal confusion about the location of Waugal. The project managers argued that the natural features identified by Aborigines as associated with Waugal also existed outside of the redevelopment area – there was another spring, another Waugal 'hole' in the river, another defile in the limestone escarpment, another tree. The development managers did not deny the presence of 'the mythical Aboriginal serpent' in the area, but sought to demonstrate that its geography was different to that 'currently believed by the Aboriginal community' and was in fact located a convenient distance east of the proposed redevelopment site (Gregson 1989: 1). Their report matched careful, rational deduction based on historical reconstruction against what was described as the 'current dis-orientation' of Aborigines (ibid.: 24).

The Old Swan Brewery controversy was not simply about an intrinsically unmappable indigenous knowledge confronting a map-addicted planning mentality. I do not want to stage what Paul Gilroy (1993: 48) describes as a 'confrontation between the regional values of a distinct sector or community and the supposed universalism of occidental rationality', to propose simply that Aboriginal lifeworlds are 'incommensurate with' those of the coloniser.

Indeed Aboriginal protesters also made use of cartographic translations of the land to make their case for its sacred status. In responding to a map produced by the Museum which failed to include a spring associated with Waugal, Aboriginal protesters referred to a 1895 plan of the original Brewery on which a 'well' was marked (Vinnicombe 1989). The protesters presumed that the well would have been built over an existing spring on the site and that, in accordance with their understanding of the Waugal Dreaming, it must indicate the presence of Waugal. The Museum argued that the well was not built on an original spring and therefore could not be seen as evidence of the presence of Waugal (*Perth Sunday Times* 5 August 1990). Contemporary Aboriginal knowledge of country is not pristinely separate from non-Aboriginal knowledge. The formation and articulation of Aboriginal land interests are not tied to some pre-contact 'truth' of an innocent and unstrategic Dreaming. The 'true', possibly unstrategic, Waugal is unknowable. Most of what is known about the Waugal (by both Aborigines and non-Aborigines) 'has been learned precisely ... in the context of specific historical relations of white–black interaction' (Merlan 1991: 345). This includes, of course, coming to know the Waugal by maps not of Aboriginal making.

BACK TO NATURE

In Robert Bropho's (1980: 16) account of his life as a fringedweller he proposes that Aborigines were 'the structure of many buildings ... all over this continent ... the foundation stone for many churches ... many cattle stations'. Bropho is of course reminding us that colonial Australia was quite literally built on the back of Aboriginal land, labour and knowledge. Historically speaking, building(s) has(have) been antagonistic to Bropho. When he and his fellow protesters confronted the plans to redevelop the Old Swan Brewery they had little time for its bricks and mortar or the measures to ensure its protection as a heritage object. Any work on the site, even the plans to undertake temporary roof work to protect the buildings, was considered to interfere with the well-being of the site and disrupt the 'natural, gentle sacred' flow of water onto the ground (Fringe Dwellers of the Swan Valley Inc. 1990: 2). The Aboriginal protesters not only opposed redevelopment, but wanted the buildings on the site demolished and the area turned into parkland for use by all residents of Perth. Bropho's (re)turn to Nature emphatically rid the site of existing and planned monuments to non-Aboriginal occupation. But even Bropho did not imagine that there was a primordial Nature to which he and his land could return. Rather, during the time he and his fellow protesters occupied the site, they set about actively (re)turning the site to Nature by planting trees and scattering seeds. In Bropho's view a gesture that was 'against architecture' and 'for nature' was the only appropriate means of a present 'remembrance and commemoration' of Aboriginal prior occupation of Perth (Bropho 1992).

Although a survey taken at the time of the controversy suggested a majority of Perth residents also thought the land should be turned into a park, this was simply not an option entertained by the government in its development aspirations for the site. A return to Nature, as proposed by the Aboriginal protesters, was deeply antithetical to the city, against the 'logic and inevitability of modernity and progress' which at this moment was articulated through consumption-linked heritage industries (Fielder 1991: 35). Of course, the city has always needed Nature, or more precisely Nature realised as resource. The making of cities, the very process of urbanisation, is not confined to the urban at all and draws on its country background, its hinterland (Pudup 1994: 116, after Cronon 1991). But, as Raymond Williams (1985) argues, this economic intimacy has also required an imaginative distantiation between country and city. Indeed, Margaret Fitzsimmons (1989: 108) suggests that 'Nature as we know it was invented in the differentiation of city and countryside.' Nature came into the city but did so as commodity (food) or as domesticated spectacle (the city park, the botanical garden or the suburban yard). The (re)turn to Nature proposed by the Fringedwellers would, of course, always be a version of such domesticated accommodations.[14]

The Aboriginal proposal for the Brewery site envisaged a conciliatory solution in which the land, once returned to park, would be opened to public use. That is, they wanted to remove the land from private property and entitlement systems. This is not an unfamiliar option in contemporary Australia. Many significant tracts of land are returned to Aborigines under land rights provisions *only* if they are then re-leased to National Park authorities for use by the public. Even under the more generous Native Title provisions, the land available for claim is usually not private property but rather already designated National Park (Baker and Mutitjulu Community 1992; Birch 1992; Rowse 1992). Increasingly, Aboriginal land rights successes are realised through options already carved out by environmental preservation objectives. Such environmentally sanctioned solutions are most readily and almost always realised in non-urban areas. The Aboriginal aspirations for the Brewery site, although far from unfamiliar to the nation, were indeed strange for urban space. A public Nature was deeply antithetical to the accumulative logic of the city. Aboriginal claims were construed as not simply 'unproductive', but also a 'wasteful' defilement of urban space (Fielder 1991: 35–36).

The government and development interests were against a return to

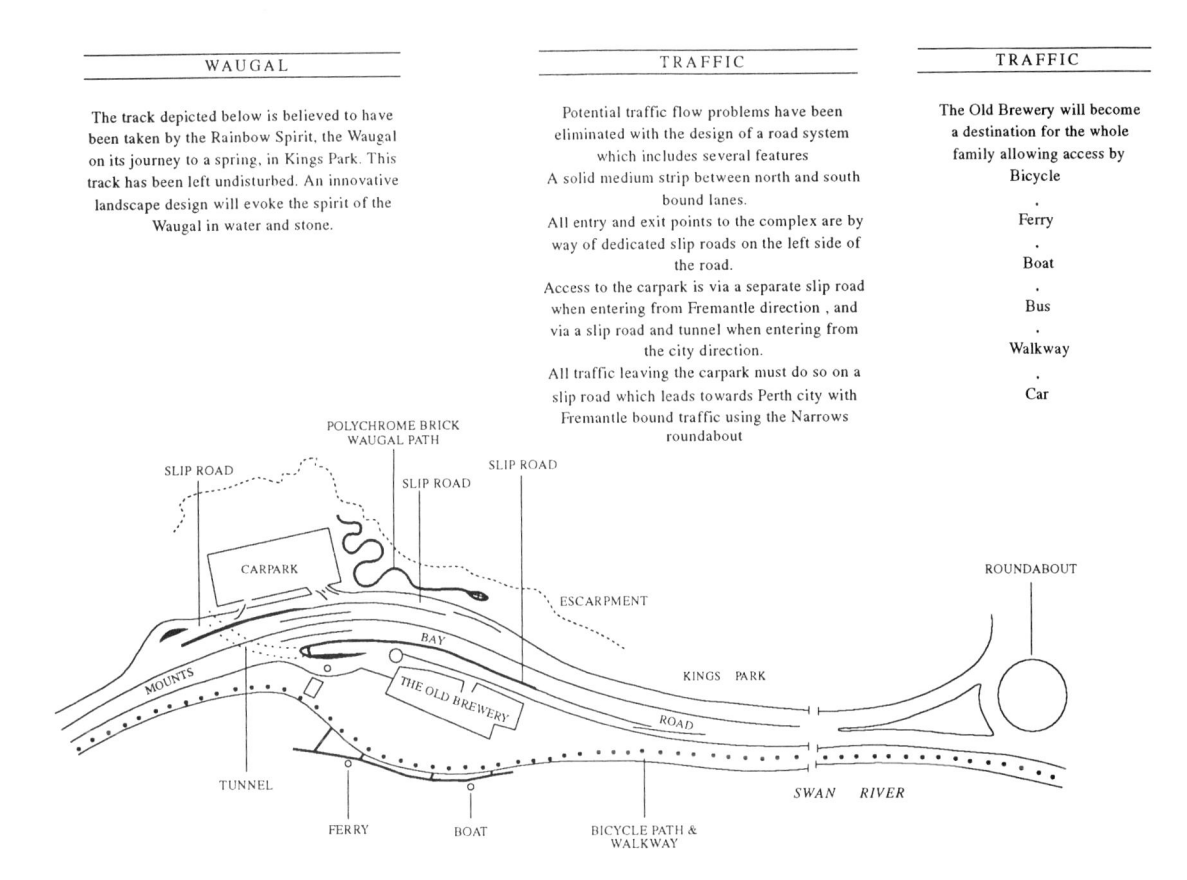

WAUGAL	TRAFFIC	TRAFFIC
The track depicted below is believed to have been taken by the Rainbow Spirit, the Waugal on its journey to a spring, in Kings Park. This track has been left undisturbed. An innovative landscape design will evoke the spirit of the Waugal in water and stone.	Potential traffic flow problems have been eliminated with the design of a road system which includes several features A solid medium strip between north and south bound lanes. All entry and exit points to the complex are by way of dedicated slip roads on the left side of the road. Access to the carpark is via a separate slip road when entering from Fremantle direction , and via a slip road and tunnel when entering from the city direction. All traffic leaving the carpark must do so on a slip road which leads towards Perth city with Fremantle bound traffic using the Narrows roundabout	The Old Brewery will become a destination for the whole family allowing access by Bicycle · Ferry · Boat · Bus · Walkway · Car

Nature which, in this case, was also a return to Aboriginal aspirations. They were not, however, opposed to the Brewery redevelopment incorporating less antagonistic markers of the Aboriginal interest in the site. The government proposed a cultural/tourism facility that would incorporate more 'appropriate' displays of Aboriginal culture. In 1988 it suggested that the Louis Allen Collection of Aboriginal Art, purchased off the US market for A$1.5 million by the Western Australian Development Corporation, be housed in a museum or gallery space in the soon to be restored Brewery. Both the State's purchase of the private collection and the gallery plans were publicised as reconciliatory gestures of cultural repatriation in the very year that marked the bicentennial of the arrival of the First Fleet on the east coast of Australia and the commencement of formal colonisation. Aborigines concerned with the protection of the Waugal site refused the 'gift'.

Later schemes for the redevelopment of the site were also eager to incorporate Aboriginal culture. Landscaping was proposed which 'evoke[d] the spirit of the Waugal in water and stone' and included a polychrome brick Waugal path (Gregson 1989; Figure 26). A museum or gallery space was again

FIGURE 26 The Roger Gregson plans for the Old Swan Brewery redevelopment gestured to the Aboriginal concerns over this site by including a 'polychrome brick Waugal path'. The symbolic incorporation of Aboriginality into such tourist ventures creates a complex politics (see Chapter 6). (Source: Gregson, n.d.)

proposed to allow the display of other Aboriginal artworks and cultural displays. The WA Museum endorsed these Aboriginalised development plans, arguing that they provided an important opportunity to 'preserve and promote respect for the cultural heritage' (*The West Australian* 30 September 1989). The development of a 'major tourist attraction' with an in-built Aboriginal theme was considered by officials as the only way to 'overcome ... the historical and spiritual significance of the site' (ibid.: 22 June 1988). But what for the government and its various agents was a reconciliatory cohabitation of tourism and marketable Aboriginality was, for Aborigines, a familiar displacement of their interests over land. From the government's point of view, if there were to be a gesture of 'reconciliation' or commemoration at this site it had to be by way of a display of commodified Aboriginal culture and not the hidden culture of a public Nature. Aboriginal aspirations to return the site to Nature precluded the expression of 'reconciliation' through marketable signifiers of Aboriginality, an explicitly cultural Aboriginality, which could be incorporated into a commodified museum system of art/culture.

PRESERVING THE CROWN

In April 1989 Aborigines protesting against the Old Swan Brewery redevelopment petitioned Queen Elizabeth II. The petition was sent wrapped in the bark of a paperbark tree, tied with string decorated with beads in the red, black and yellow colours of the Aboriginal flag. She did not respond (Ansara 1989: 37). This was just one of many appeals made by Aboriginal protesters to and against 'the Crown'. The precariousness of 'post'coloniality in nations like Australia is clearly expressed in the technologies of power contained in the law.

The conflict over the Brewery became a test of whose 'heritage' would be given the privilege of the legal protection that both were afforded under law. The Waugal sites at the Brewery were officially recorded in 1985 and, under the provisions of the Western Australian Aboriginal Heritage Act, the developers would normally be obliged to consult with the Museum prior to the commencement of on-site work. The Western Australian Development Corporation failed to do this and subsequently disturbed Waugal sites. The basis of the Aboriginal legal challenges to the development was that the State of Western Australia, through the Western Australian Development Corporation, had breached its own law. In its court defences, the WADC argued that the Aboriginal Heritage Act applied only to citizens of the Crown who were owners of land leased from the Crown, but not to *the Crown itself*. In short, the government proposed that because the WADC was an agent of the Crown it was exempt from the Act. Aborigines took this case to the courts three times in the course of their protests.

Challenges made in the State Supreme Court found in favour of the government's development agency on the basis that the Crown and agents of the Crown were conventionally exempt from the regulatory frame of statutory provisions. Historically this provision was intended to protect the privileged rights, the majesty, of the actual person of the Sovereign; in the current case, Queen Elizabeth II. It is an assumption that law is made 'by rulers for subjects' and is not intended to apply to 'the Crown' (Churches 1990: 7). In practice this assumption is not confined to the Sovereign herself but extends to confer prima facie immunity in relation to the activities of governmental instrumentalities or agents of the Crown acting in the course of their functions or duties. In the setting of 1980s Western Australia, the activities of the government reached into almost all aspects of commerce, industry and development and the Crown regularly competed with and had commercial dealings on the same basis as private enterprise (ibid.). The privilege of the Crown extended well beyond 'the Sovereign' and worked to protect a state that was operating like 'a subject'. And as some 90 per cent of land in Western Australia is in the hands of the Crown, of which only 40 per cent is leased to private interests, this legal loyalty to the privilege of the Crown provided the basis for substantial exemptions.

The failure of the WA Museum, the agent with the legal responsibility to protect Aboriginal sites, to advise against redevelopment provided the basis for Aboriginal protesters once again to take their case to the courts and argue that an agent of the State did not abide by its own statutes. In the High Court finding the judiciary unanimously agreed that the Crown was not immune from its own laws and that the Western Australian Government's agencies had broken the law by not protecting Aboriginal sites from development. The finding overturned earlier legal precedent as well as any adherence to recent decisions under British law which preserved the privileged rights of 'the Crown'. As one legal commentator noted, the decision 'marks a very clear divergence between British and Australian attitudes to the position of government's relationship to the law' (ibid.). The decision was seen to have enormous legal ramifications and to be deeply unsettling to legal arrangements 'which have been in place and stable for many years' (WA Attorney-General quoted in *The West Australian* 21 June 1990). Within a month of the High Court decision the Western Australian Parliament had introduced additional legislation to clarify the position of the Crown in future development issues.

The Aboriginal protest did not result in the saving of the Waugal ground. The Western Australian Government was still able to exercise its discretionary powers and proceeded with development for 'the benefit of the whole community'. The preservation of buildings that were considered 'a national heritage asset' won out over the minority interests of Aborigines (Minister for Aboriginal Affairs, quoted in Churches 1992: 9). The Perth planning and development system was not incapable of registering an Aboriginal interest in the land, but could do so only in ways that gave a marketable indigenous 'culture' to development visions. The Aboriginal sacred had once again been 'properly' placed in the urban scene.

Yet the triumph of the city was not that of the nation. The legal victory of the protesters testifies to the force of the Aboriginal sacred in shaking up the nation. In the legal battles around the Swan Brewery redevelopment the Waugal worked to rearrange radically one fundamental residue of colonialism in Australia, the legal privilege of 'the Crown'. Colonialism has required a repression of the Aboriginal sacred and actively and anxiously continues this process. But in doing so it actually helps to produce/reproduce the Aboriginal sacred, to amplify its presence, to ensure that it insinuates itself into the making of the nation.

FRINGEDWELLING

'The whiteman, he believe it's just the blacks sitting there in big mobs in tents singing out, "Sacred site! Sacred site!"; this is what he wants to believe. This is why he sings out to us as he's passing by "Go home you black 'Cs!" "Go home you black 'Bs!". We can't go home – because we already home.'

(Bropho, in Bropho and Ansara 1989)

From January to October 1989 the 'invisible' Aboriginal residents of Perth camped at the Old Swan Brewery (Figures 27 and 28). They dwelled there: cooked and ate, slept, planted trees and went about the business of organising their campaign. The ten-month occupation of the site by Aboriginal protesters and their sympathisers was a radical disruption of the spatial ordering of the city. While many Perth residents may have thought the claims about the Waugal were imaginary, belonging to the realm of myth, they could no longer deny that there was an Aboriginal citizenry in the city. The on-site protest was a reterritorialisation of urban space which defied the measures to repress and contain Aboriginality in the city. The Old Swan Brewery site had been incorporated into the urban as a transparent space: it was 'public' space, a visual landmark, thoroughly mapped and zoned. These processes sought to eradicate 'the domain of myth . . . the irrational' and set the site surely within the development logic of the city (Vidler 1992: 168). The Aboriginal occupation confirmed the presence of something previously unknown and, as the futile flurry of mappings suggested, possibly unknowable. Aboriginal occupation and articulations of the presence of the Waugal transformed transparent space into opaque space and unsettled the ordered zoning of the city.

The occupation transgressed the network of closed frontiers which framed city space and through which the division between colonial Self and Aboriginal Other was secured and managed (Lefebvre 1991 :175–176). Within the conventional spatial ordering of Perth, the Aboriginal presence at the site was 'illegitimate'. Aboriginal claims to be 'caring for their country' were challenged by media coverage which reported that Aborigines were defecating in a tunnel which runs under the road. Fielder, drawing on Mary Douglas's (1984) understanding of the relationship between purity and pollution/order and disorder, argues that these charges of defilement depicted the Aboriginal occupation as 'impure', and returned the Aboriginal dwellers to a more familiar stereotype of a 'lesser human being – inherently degenerate, lazy, unreliable, dirty and disorderly' (Fielder 1991: 36).

For the protesters the act of reterritorialisation was an opportunity to affirm meaning, not only in the wider sphere, but also for themselves:

> When we come to this area here there was nothing here. Since then, and now, a lot of days and nights and weeks and months have gone past. The whole of this area is all sacred to the Aboriginal people.
> (Bropho, in Bropho and Ansara 1989)

The occupation of the site became the moment at which the Fringedwellers reactivated a (pre-)modern knowing of space within the specific conditions of modernity. This was an occupation that helped to produce meaning. But here modernity is not interested in reactivated meaning or in Aboriginal reterritorialisations. When adjudicating on the validity of Robert Bropho's claims to the Brewery site, officials argued that his interests were 'emotional and intellectual' and not linked to 'a physical association with the site'. Bropho had not 'habitually camped there', he had not 'hunt[ed] or fish[ed] or take[n] children there to teach them culture, or [held] ceremonies there, or otherwise frequent[ed] the area' (Justice Anderson, cited in Churches 1992: 10). Despite Bropho's claims to the site, despite his leading hundreds of Aborigines to reterritorialise this land, he was cast back into a form of fringedwelling by non-Aboriginal adjudications of what was proper Aboriginal dwelling.

FIGURE 27 (*over page*) Robert Bropho (left) and a fellow protester at the Old Swan Brewery site. Bropho and the other protesters wanted the buildings demolished and the area (re)turned to parkland. Their desire to return the land to 'unproductive' Nature was as disruptive to city development aspirations as the previously unknown presence of Waugal. (Reproduced by courtesy of *The West Australian*)

FIGURE 28 (*over page*) As soon as the Aboriginal claims were overruled by the courts the police moved in to remove protesters. The protesters had been at the site for many months and their experience of dwelling at this place itself reactivated the meaning and value of the site. Their presence at the Brewery disrupted the view that Perth had effectively rid itself of Aborigines. (Reproduced by courtesy of *The West Australian*)

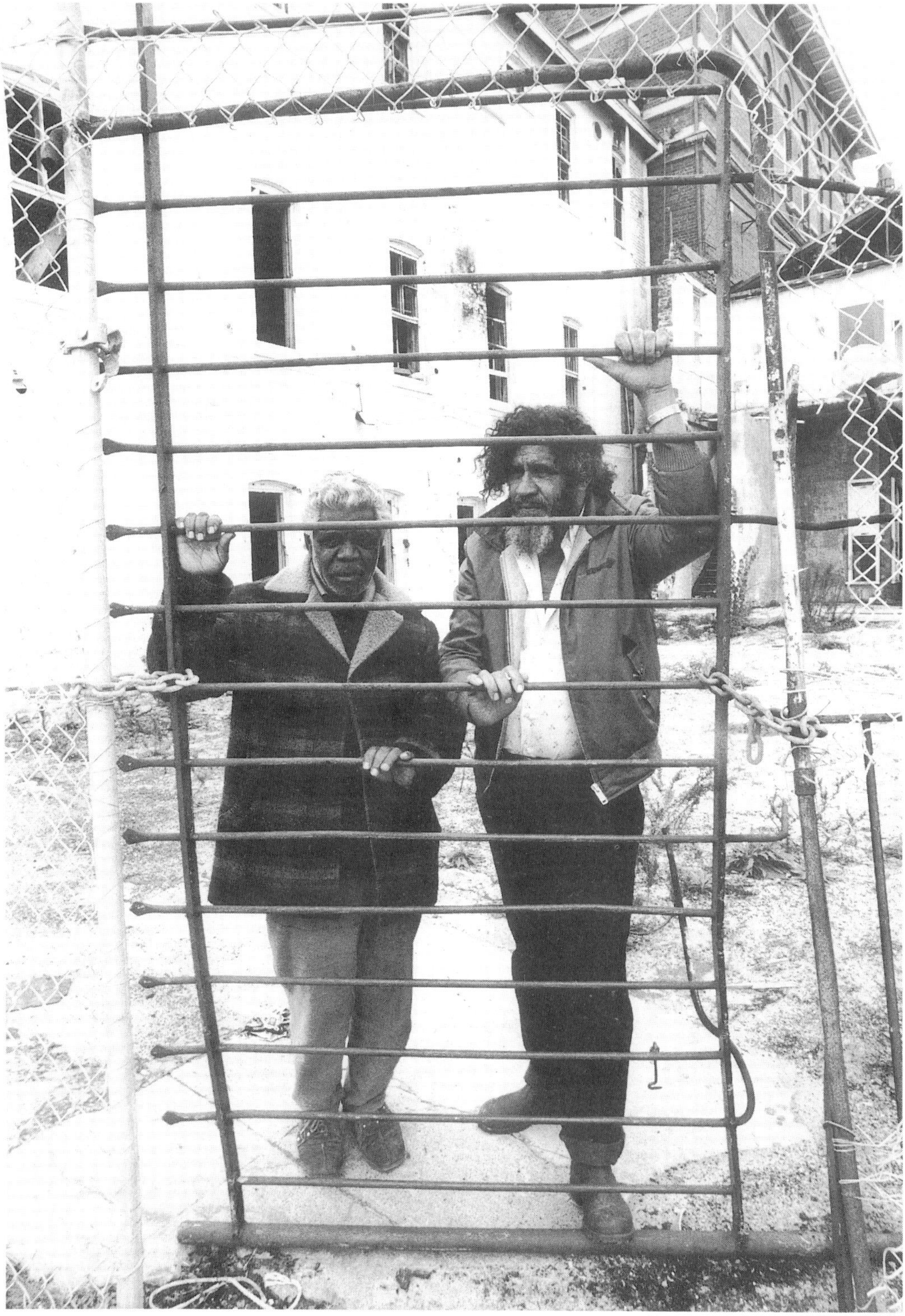

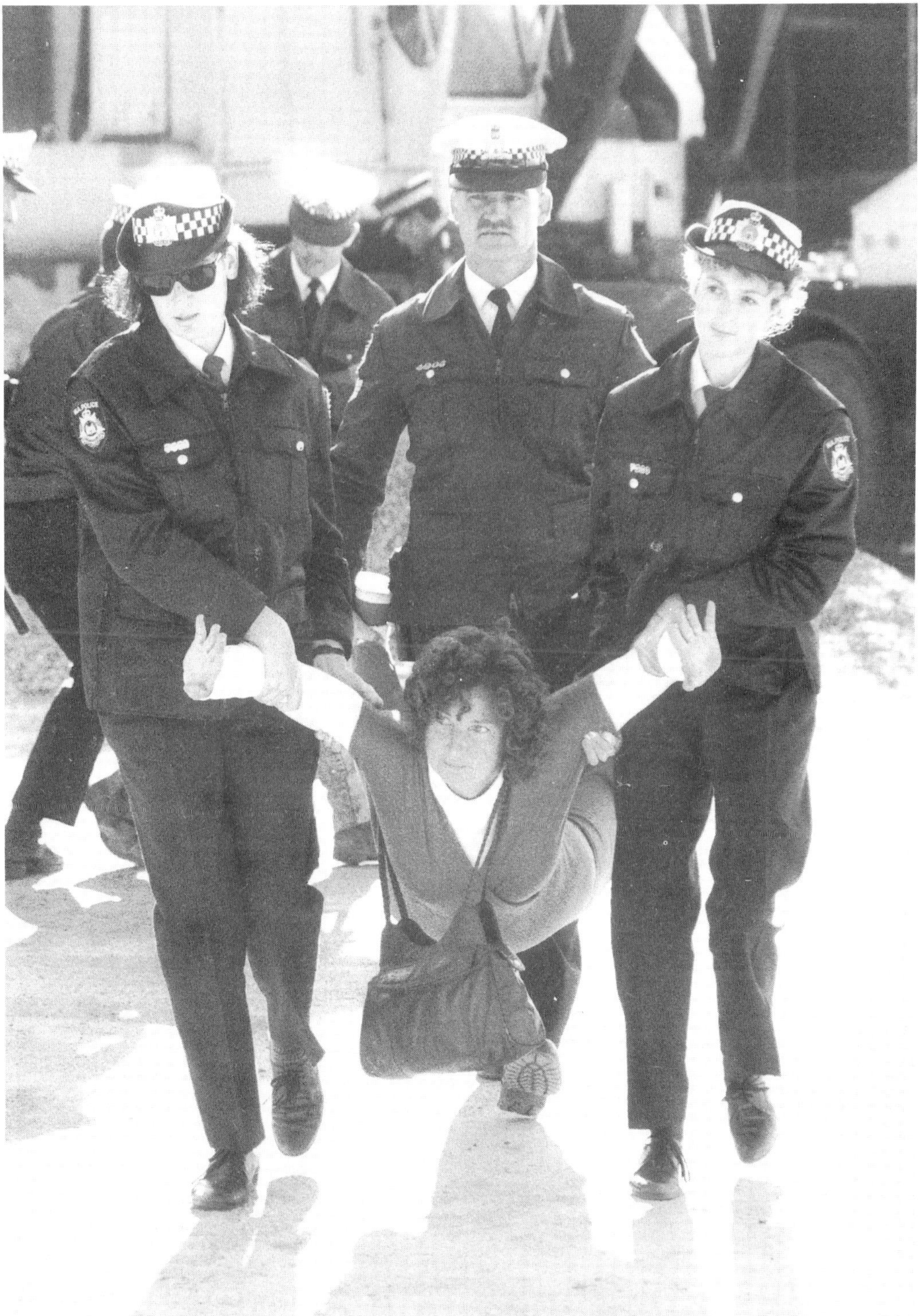

Urban development in colonial Australia attempted to locate Aboriginality outside of the city. This was a spatial repression which kept distant the Otherness against which the Self was constituted. The Swan Brewery controversy marked an uncanny return of this displaced otherness. The homely city became 'unhomely', was rendered unfamiliar to itself by a transgression of the 'fragile boundaries' which marked that 'uncertain Self' (Kristeva 1991: 183 and 188). This uncanny return is the failure of colonial mastery over the 'Other'. As Bhabha (1985) suggests, this mastery is never fully realised, it is always slipping, always being reinscribed and always producing an unmanageable excess. The uncanny appearance of an embodied Aboriginal occupation and an unknowable Aboriginal sacred in the secular space of the city of Perth set in train an anxious politics of reterritorialisation. Aboriginality was no longer emphatically placed 'outside' of the city and non-Aboriginal interests anxiously sought to ensure that it came 'inside' in manageable and marketable forms. While Aboriginal protests 'failed' in the sense that the development proceeded, they also reactivated, however fleetingly, Aboriginal knowledge and occupations of urban land. Indeed, the neo-colonial attempts to repress the Waugal produced the very conditions by which it could escape its place on the banks of the Swan River and unleash its disturbing powers on the constitution of the nation.

NOTES

1. 'Dreaming' is the popular term used in Australia to describe the spiritual and ritual life of Aborigines. It is often used more specifically to describe the time when creation beings moved across the earth forming waterholes and other natural features. I also use 'dreamings' (without a capital) in a more general sense to refer to the visionings of developers, etc.
2. The Waugal lives on the high ground, where the State Parliament House now stands. As he travelled across the country he created the Swan River and other features on the land. The site of the Old Swan Brewery was one of Waugal's resting places.
3. Mickler (1991a and b) has produced a nuanced spatial reading of the struggle over the Swan Brewery site. His 'cartography of the tribal and the modern' (Mickler 1991b: 71) opened out many cogent themes associated with this case study. My indebtedness to this pre-existing reading is evident in the text and my return to this event is in the spirit of extending and elaborating Mickler's analysis within the broader concerns of this book.
4. The various deals and collusions which mark this period of laissez-faire development culminated in a Royal Commission into Commercial Activities of the Government and Other Matters, popularly referred to as the 'WA Inc. Royal

Commission'. The state Labor leader of the time, Brian Bourke, like many of his big businessmen partners, is currently engaged in a lengthy post-mortem conducted in courtrooms and gaol cells.

5   Designs for the redevelopment were commissioned from the architect Roger Gregson.
6   The passage of the Native Title Act did result in a number of Aboriginal groups making what were called 'ambit' (unproven) claims over urban lands.
7   Studies of the postmodern city, such as those by Ed Soja (1989) and David Harvey (1989), speak of a surprisingly secular city. This erasure of the sacred speaks precisely of the modernist underpinnings of 'western' urbanism as well as the recalcitrant modernism of these accounts of the 'post'modern urban.
8   Some government agencies have been successfully lobbied to ensure that 'secret sacred registers', which may only be accessed with the permission of traditional custodians, are established.
9   That the Museum was the chosen authority to administer the procedures for site registration attests to the close connection between such recoveries and earlier colonial practices of collecting and categorising Otherness.
10 Aborigines had been disclosing the presence of the Waugal at this site for some time.
11 Statutory protection for the historic built environment was only available through recourse to Federal provisions developed through the Australian Heritage Commission Act of 1975.
12 In 1990 the Heritage of Western Australia Act was passed and a Heritage Council established. This Act was to provide for the 'conservation of places which have significance to the cultural heritage of the State'.
13 In this initial assessment of the site there was some doubt expressed about the historical integrity of the Brewery buildings, which were a combination of the original 1897 structure and additions made during the 1920s and 1930s. By the assessment of some, the buildings were an urban ugliness rather than a heritage asset (Western Australia House of Representatives 1991: 41220).
14 Indeed, as I show in Chapter 6, the contemporary city is increasingly disposed towards a (re)turn to Nature.

6

AUTHENTICALLY YOURS

DE-TOURING THE MAP

•

maps are preeminently the language of power not of protest.
(Harley 1988: 301)

In 1994 the local planning authority for the eastern Australian city of Brisbane in the State of Queensland issued a development brief for a A$10 million ecotourism centre. The centre will focus on the somewhat unlikely assemblage of Aboriginal culture and what is claimed to be Australia's national symbol, the koala (Brisbane City Council 1994a: 1). Here 'native animal attractions' and 'Aboriginal cultural attractions' will be unified in what the Council itself described as 'a marketable image of how the Aboriginal culture has co-existed with the natural environment for several thousand years' (Brisbane City Council 1994b: 26). The centre seeks to provide a journey back to an Australia prior to white settlement when Aborigines were 'in harmony with nature and [their] surroundings'. It is estimated that once completed in 1996, between 197,000 and 244,000 people will visit the centre each year (ibid.: 15). Queensland is a key tourist destination for both domestic and international tourists, in part because of the climate but also because of attractions like the World Heritage listed Great Barrier Reef. The State captures 23 per cent of the national tourist market and some 47 per cent of international visitors spend at least part of their visit to Australia in Queensland. Between 1987 and 1994 there was a 4 per cent growth in domestic tourism and a 21.8 per cent growth in international visitors to the State (Laurance 1994: 6). Tourism is considered to be the State's third largest industry and injects A$4.4 billion annually into the economy and provides a key source of employment (Queensland Tourist and Travel Corporation 1992: 28). As State capital and the major transport interchange, Brisbane benefits considerably from the State's tourism industry and in recent years has begun to plan future development more fully around tourism (Brisbane City Council 1993a). 'New' tourisms, such as cultural tourism and ecotourism, are seen as crucially important in future development. Indeed, the plan for an ecotourism centre was held to be the flagship of the city's intention to enter wholeheartedly into an economically viable future built on tourism.

A recent survey of tourism in Australia found that some 49 per cent of international tourists coming to the country are interested in experiencing Aboriginal culture and that one in five visitors to Australia goes to a museum or a gallery specifically to view Aboriginal art (Australia Council 1990). The Brisbane ecotourism centre seeks to capture this heightened interest among tourists in Aboriginal culture. Aboriginality is to be incorporated into the proposed ecotourism development from the foundations up (Figure 29). The paved floors of the centre will include mosaics inspired by Aboriginal designs and the colour scheme will be based on an 'authentic' Aboriginal palette. The glass walls are to be sand-blasted with Aboriginal figures which will hang 'ghost like' between the ecologically sound, pre-modern Nature realised inside and the defiled Nature which exists in the modernity outside (Brisbane City Council 1994b: 36). Visitors initially will be led to a large area designed after 'an ancient cave' and covered in traditional rock art designs. Here they will encounter a large array of interactive displays which show various aspects of the Aboriginal relationship with the land in pre-contact times. Visitors will then be led outside to an amphitheatre where they will watch Aborigines performing traditional dances or making traditional artefacts. Alternatively visitors might accompany an Aboriginal guide on a 'treetop tour' to hear about Aboriginal food and medicinal uses of the bush. They might even have their

FIGURE 29 The Brisbane City Council brief for the ecotourism centre proposes a building which is 'at one' with Nature. The building is to incorporate Aboriginality from the foundations up. (Source: Brisbane City Council 1996b)

bodies painted in traditional Aboriginal designs or try their hand at playing the didgeridoo or throwing a boomerang. Weary from their immersion into the ecoculture of pre-contact Australia, visitors might then sit down to an Aboriginal-inspired 'bush tucker' meal in the on-site restaurant or browse amongst the 'high quality' artefacts available for purchase in the visitor shop (ibid.: 16–17). It is anticipated that Aboriginal contributions to the centre will need to be adjusted and modified – made more 'clearly spiritual', more 'technically advanced', more 'comic' – in order that they might 'suit the needs of the market [and] ensure visitor expectations [are] met' (ibid.: 16 and 27). In short, the ecocentre requires not just any old traditional Aboriginal practice, but a 'professional' traditionalism, honed to the requirements of the ecological spectacle.

It is perhaps not surprising that those proposing this highly contrived and excessively commodified proposal to market Nature and Aboriginality in harmonious unison have displayed a certain anxiety about authenticity. In the Aboriginal displays, performances and products 'authenticity' was designated as a 'crucial element' in the success of the venture and any failure to evoke this necessary ingredient would, it was proposed, detract from visitor satisfaction (ibid.: 14). Perhaps ironically, this concern has not been generated by any sense of the internal inevitability of a displaced authenticity, what Morris (1988a: 13) refers to as tourism's own ability to 'ambush' the very authenticity it treasures. Rather, the concern for authenticity is driven by a belief that Australia is already littered with too many 'commercialised', 'glossy' and 'non-authentic' versions of Aboriginal culture. It is, of course, through the proliferation of what might be seen as not original that 'the eminently modern value', the fantasy of 'authenticity', is marked out (Frow 1991: 129). A patina of authenticity is a way of marking this ecotourism centre out from the many other attractions already trading in Aboriginality: a way of making the centre not only a more desirable tourist adventure but also, of course, a more viable economic venture (Brisbane City Council 1994b: 14).

To ensure the necessary level of authenticity it has been proposed that Aborigines be involved in both the initial planning stages of the ecocentre and the ongoing presentation and performance of the 'eco' spectacles. This planned participation would, it was hoped, produce a cosy union between morality and money:

> The attraction must be perceived as being ethical eco-tourism, that is authentic, with both the natural environment and the Aboriginal people benefiting from the establishment of the attraction. This can then be used as a marketing tool in encouraging patronage of the park.
> (ibid.: 26)

It is undeniable that non-Aboriginal Australians seeking to develop some form of meaningful co-existence with indigenous people need to engage in both moral and practical reasoning (Rowse 1993: 229). But here it is hard to discern where a moral practice ends and a theatre of ethics begins. In that it is recommended that Aboriginal participation stops short of economic control of the centre, because of their perceived 'lack of business experience' in this area, it seems that in this venture the spectacle triumphs (Brisbane City Council 1994b: 28).

Brisbane's proposed ecotourism centre is just a vision and if, as proposed, development proceeds in consultation with Aborigines, it will no doubt undergo many adjustments. As an as yet uncompromised vision it provides a stark example of many of the issues raised by the emergence of 'new' forms of tourism which are being labelled as 'ecotourism'. Ecotourism is premised on the notion that there is now a new desire among tourists to visit sights that are either 'of Nature', an unspoilt 'wilderness', or 'about Nature', more precisely, about how humanity can better be in the natural world. Mostly it is a form of anti-tourism which consciously promotes the experience of encountering places 'unspoiled' either by mass tourism or other markers of modernity. Signs of human habitation are not precluded from the 'Nature' sought out by the ecotourist. The heritage building and Aboriginal cultural sites are frequently absorbed as resources in this new, niche-marketed tourist industry. Indeed a recent Australian study (Brokensha and Guldberg 1992) sees ecotourists and cultural tourists as comparable consumers, seeking similar experiences and requiring similar management strategies.

NATURE, CULTURE, COLONIALISM

At the most fundamental levels, Nature, including its primordiality, is a socially mediated and constructed notion. Nature is also, as Raymond Williams (1980) argues, deeply imbricated in the material and affective practices of society. Part of the legacy of the cocktail of Enlightenment thinking and the transition to capitalism was the invention of 'external', 'primordial' Nature. As Margaret Fitzsimmons (1989: 109) notes, this notion of an 'externalised, abstracted, Nature-made-primordial provides a source of authority to a whole language of domination', including of course that which Marx expanded – the capitalist domination not just of labour but also Nature (see N. Smith 1984; Cronon 1991). The ecotourism industry is a deceptive elaboration of these processes of construction and domination. Here the (re)turns to Nature pursue either a fantasy of the primordial, Nature untouched by culture, or, as in the case of Brisbane's ecotourism centre, a spectacle of Nature properly touched by culture. These are (re)turns that displace or reform culture's domination of Nature; they market what Donna Haraway (1989: 264) calls a 'politics of healing'. But of course, at the same time, they mark an adaptation of culture's (capitalism's) appropriation and mediation of Nature. The various pathways by which Nature is self-consciously met in the modern world – conservation, sustainability, ecotourism, environmentalism – are in this sense underpinned by what Haraway (ibid.: 265) refers to as a 'dialectic of love and money'.

It is not Nature that is on display at the Brisbane ecocentre, but a culture that knows Nature differently and, presumably, in a less destructive form. Nature is found by way of Aboriginal culture. Jennifer Craik (1994: 153) argues that tourism ventures based on re-presentations of indigenous cultures establish certain sets of relations, such as voyeurism and extractive commodification, which 'mimic colonialism'. Indeed, there is a growing acceptance that consumption-based industries like tourism are the face of contemporary colonialism, a form of neo-colonialism (Boniface and Fowler 1993: 19–20). Certainly the plans for the Brisbane ecotourism centre reiterate a number of familiar colonialist relations. This is evident in the exclusion of Aborigines from the economic management (and full economic benefit) of the Centre. It is also apparent in the very way Aboriginality itself is packaged. It is a pre-contact, primitivised Aboriginality that is required in the ecotourism centre, albeit one that is adequately adjusted for entertainment value. There is an undeniable resonance between this eco-sensitive display of Aboriginality and nineteenth-century practices of framing Aboriginal culture as primitivised 'spectacle' – in museum spaces, in anthropological systems of knowledge and collections of exotica (von Sturmer 1989: 139).

The ecotourism centre represents a commodified version of a more general and variably expressed modern desire to (re)turn to Nature by way of indigenous cultures, to see indigenous peoples as the First Conservationists. That is, the affective regimes at work in the idea for the ecocentre resonate with a much broader field of desire articulated in a wide range of environmentalist formations. This is expressed within some strands of the environmental movement, and in particular Deep Ecology (for example, Drengson 1989; W. Fox 1990; Knudston and Suzuki 1992), and in ecofeminist environmentalisms which draw not only on indigenous cultures but also on certain models of 'woman' within pre-patriarchal societies (for example, Griffin 1978; Merchant 1980, 1992). Indigenous culture also appears in a variety of other environmentalist expressions: offering the inspiration for the marketing of environmentally sound products (Sackett 1991); providing templates for New Age ecospiritualisms (for example, M. Fox 1991; Lawlor 1991), and models for land management in nature conservation areas (Baker and Mutitjulu Community 1992). The (re)turn to Nature by way of indigenous knowledges is a response to what is seen to be the failure of masculinist and rationalist (read capitalist) ways of seeing, knowing and being in the world. For example, Knudston and Suzuki's *Wisdom of the Elders* explicitly turns for guidance to 'Native peoples' intellectual and experiential insights into "proper human relationships with the natural world"' (1992: xiii–xiv). The appeal of indigenous knowledges is what is recategorised as their 'ecocentric' perspective: the ways in which land-based indigenous systems of knowledge integrate culture and Nature and do not prioritise the human above other species and things. The radical subjectivity, or 'transpersonality', of these systems of knowledge is seen to provide templates for the modern development of a wider sense of self, what Mathews (1992) calls an 'ecological self', which includes all beings and all things.

In these environmental reimaginings of the ecoplanet there is, not surprisingly, little sense of boundary – traditional Aboriginal land knowledge and non-Aboriginal aspirations for a sustainable being-in-the world are seen as potentially one and the same. 'Indigenous knowledge' is seized upon as providing a cultural model for a modernity that might construct itself not around masculinised 'anthropo'centrism, but through a decentred subjectivity – 'us' as a part of, and at one with, Nature (see Jacobs 1994b). Indigenous peoples are placed back in First Nature, they have become the First People of the world, and then are reabsorbed into a global chronology of planetary survival, which begins with 'them' but ends with an environmentally sound 'us'. As Thomas (1994: 174–177) points out, this new ecological imagination draws on primitivised, stable, ahistorical and deeply romanticised understandings of Aboriginality and Aboriginal associations with the land. In this variously expressed politics of sympathy, 'traditional Aborigines' become both a people upon which to 'project feelings about the present' and from which to draw 'blueprints for the future' (Torgovnick 1990: 244).

Viewing Aborigines as 'in harmony' with the land, at one with and part of Nature, has been a persistent trope of colonialism (Lattas 1990). Under nineteenth-century colonialism the conflation of Aborigines and a feminised Nature legitimated the placement of Aborigines as outside of civilised Culture: sometimes passive, sometimes capricious or wild, but always available for masculinist conquest (Schaffer 1988). These constructs, which articulated themselves through romanticised yearnings for a pre-modern Other and audacious disregard for indigenous rights over land, provided important preconditions for the colonial occupation of Australia. The contemporary eco-version of the Aborigine/Nature fold may have new forms but is just as able in its colonialist effects. These articulations demonstrate the necessity of an 'authentic Otherness' in modernity's quest for an ecologically sound future. Aborigines who slip outside of or choose not to engage in primitivist stereotypes of their culture have an uncertain place in this journey to a new planet. The colonialist effects of the new eco-sensibilities run further than the reactivation of an Aboriginality which may no longer fit. In their local Australian articulation, these environmentalist and ecospiritualist (re)turns to traditional Aboriginal ways of being in the world also operate as templates for how settler Australians may better dwell in the nation. As one ecospiritualist proposes, the Aboriginal Dreaming provides an 'origin' for the modern realisation of 'our root and foundations as Australians' (Cain 1991: 78). The future of the ecologically sound nation depends, then, on settler Australians making the move from seeing Aborigines as Other to themselves, 'outside you', to finding an Aboriginality '*in* you' (M. Fox 1991: 7). In this articulation, planetary environmentalism mutates into a new eco-nationalism. As Andrew Lattas (1990: 52) notes, possessing Aboriginal knowledge is not only the final step in a more ecologically sound Australian nation but also in a process of colonisation in which settler Australians can, at last, make the move from aliens to indigenes.

In the example of Brisbane's proposed ecocentre I have depicted tourism as one point in a reinvigorated colonialism. Without doubt the scale of this development and its highly commodified form contribute to its colonialist effects. The colonial problem posed by tourism may not be confined to its recent manifestations. Tourism and touring have long played a part in the way in which colonial nations like Australia were settled, known and inhabited. This is to propose that movement is a key means by which people come to dwell in place. In the past, as in the present, tourism has operated on particular regimes of desire in which Nature and Aboriginality were variously incorporated. In the remainder of this chapter I take us to a tourist site, Mount Coot-tha, which lies on the edge of the city of Brisbane and which has long provided a favoured tourist destination for local residents and visitors. Through readings of two manifestations of tourism in this place, one past and one present, I attempt to outline a more unruly politics of identity and place which may inhabit the practice of tourism.

IMPERIAL TOURING

Mount Coot-tha has long been a tourist destination for the residents of Brisbane. The mountain may well have retained an Aboriginal name ('Coot-tha' is reported to mean 'the place of honey'), but it was a colonising vision that was served by this emergent tourist space. Located some 5 miles west of the emerging city centre, and affording a vantage point across the city and surrounding areas, it attracted pleasure-seekers from the earliest days of settlement. At the turn of the century the area was described as a 'favourite ... place for picnicking' and as early as 1880 the area had been placed under Trusteeship as a reserve for public recreation. During an 1882 visit to the colony by the Duke of Clarence (then Prince Albert) and Prince George (later King George V), the Royal guests were taken to Mount Coot-tha to view a fragment of Great Britain's territorial dependency. They planted two Moreton Bay fig trees at the crest of Mount Coot-tha (Mally 1914: 3). These natural memorials marked the imperial endorsement of Mount Coot-tha as a tourist destination for local Brisbane residents and visitors to the city.

The panoptic view of the surrounding landscape offered by Mount Coot-tha was enthralling and it formed the basis of much of the early tourist promotion. From the mountain, an early tourist brochure claimed, 'one can distinguish almost every feature of the near landscape' (Traill 1902: 2). Mount Coot-tha was a place where the original gaze of colonisation could be repeatedly re-visioned. Visitors were invited to inhabit the colonial survey of the early settlers of the Brisbane region:

> the views from Mount Coot-tha [have] memories of past times.... The eager Logan and the Commandants who succeeded him have been seen on this eminence scanning the far-extending landscape, observing the numerous threads of smoke which indicated to them where the abounding natives were grouped around their fires, and pondering on the mysterious possibilities of the unknown interior.
>
> (ibid.: 16)

Early tourists to Mount Coot-tha, 'sightseers' as they were appropriately called, could imagine, as the early settlers did, a land where those 'threads of smoke' were replaced by 'painted cottages' and even riverside 'palaces' (ibid.). Visitors could step back into the colonial desire for the 'beautifying interventions' which led to the ordered scene of settlement that now lay before them (M. L. Pratt 1992: 205 and 217). As in the original colonial gaze, the turn-of-the-century vista from Mount Coot-tha was a point from which a teleology of dwelling in Australia could be imaginatively envisioned. Here the modern visitor was offered the opportunity to reimagine the movement from 'castaway to colonizer' upon which the Australian nation and their citizenship of it was built (de Certeau 1986: 143). Visitors could also imagine a most complete colonial future in which Brisbane was the imperial city of the heart transported south of the Equator. A time when:

> [t]he cold and foggy valley of the Thames may ... have lost its present throng of inhabitants, and under a kinder sky and in more genial climate, there may here be that concentration of population.
>
> (Traill 1902: 16)

Just over a decade later, Mount Coot-tha was again actively promoted as the place from which to view Brisbane and surrounds. Then, the views offered from the mountain were compared to those available from the Eiffel Tower in Paris, from the summit of Rigi in Switzerland, and from atop the Woolworth Building in New York. The 1914 promotional document assures readers that Mount Coot-tha is part of a cosmopolitan not a pre-modern nation; a place transformed from its 'wild pre-historic days' as 'the happy hunting grounds' of 'the dusky sons of the soil' and placed into an emerging global network of tourist destinations (Mally 1914: 2).

Under the guidance of its Trustees and using government funds, Mount Coot-tha underwent various on-ground 'improvements' which aimed to transform it into 'one of the most attractive places in the Southern hemisphere' (ibid.: 3). The existing steep, rough bridle tracks were replaced with well-made roads and paths so that 'pedestrians and motorists [could] attain the highest points at Mount Coot-tha with the utmost ease and comfort' (ibid.). Trees were planted to provide shade for visitors, seats and picnic sheds were built, water and 'other conveniences' were provided, and a kiosk selling light refreshments was erected. The main view from Mount Coot-tha may have been 'given' by Nature but supplementary views were literally carved out by the will of Culture. Two main vantage points were designated, and six 'large windows', or look-outs, were cut through the scrub along Coronation Drive, the main road around the mountain, in order to produce new views and vistas. At each of the new vantage points, 'finger-posts' were installed which directed the visitor's gaze to specific points of settlement, both visible and distant. The views created were ordered according to the logic of the compass placing visitors at the centre of a comprehensively named panoptic (Figure 30).

The promotional booklets for the area provided detailed verbal descriptions of what could be seen from the mountain. As a writing genre these descriptions shared much with Romantic and Victorian travel writings which aestheticised the landscape and transformed it into a verbal painting (M. L. Pratt 1992: 204). These precisely ordered and comprehensively named views worked to domesticate any traces of the boundless and terrifying potentialities of the sublime and create instead a spectacle of settlement.

> An Amphitheatre of Grandeur is spread out to sightseers as soon as they reach the kiosk on Mount Coot-tha. Standing on the stone pillar

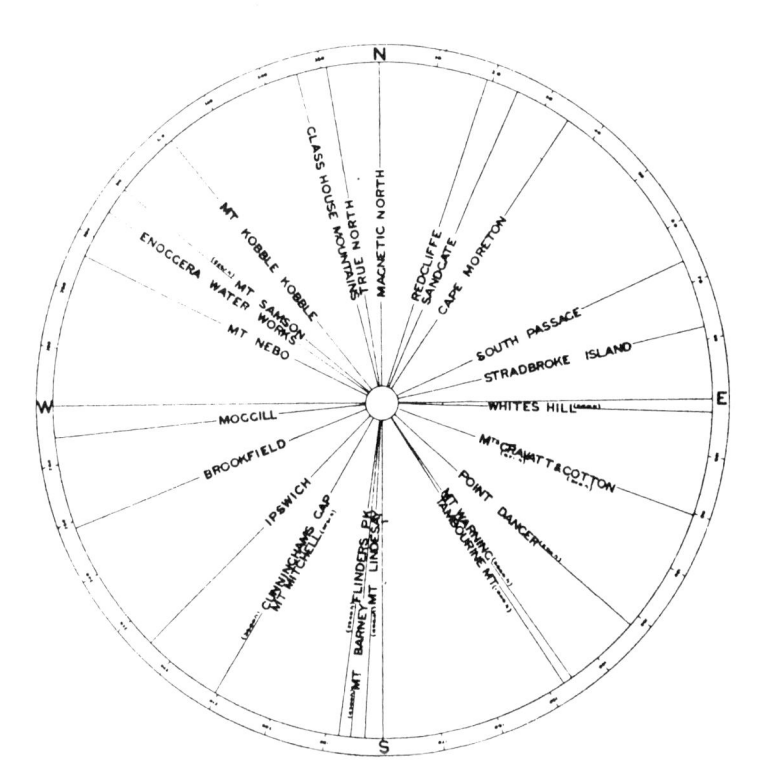

DIAL SHOWING BEARINGS FROM MOUNT COOT-THA.

(on which ... there is an engraved dial showing the geographical places of interest at the different points of the compass), it is possible, on a bright, clear day, to get a comprehensive panorama of the surrounding country in all its glamorous settings. The concatenation of views which are brought under the visual observation of the visitor are intoxicatingly entrancing in their splendour.

(Mally 1914: 6)

The comprehensive vision of the Brisbane area offered to tourists from Mount Coot-tha suggests that the mode of 'monarch-of-all-I-survey' has a considerably longer life than the initial moments of imperial visioning (M. L. Pratt 1992: 201). Unlike many of those early visionings, the on-site interventions at Mount Coot-tha and the promotional booklets were not re-inventing an imperial vision to serve a distant metropolitan core. The 'monarch-of-all-I-survey' practice created at Mount Coot-tha served settler Australians who visited the mountain. Here the imperial vision was being made into an everyday settler vision; to be had on Sunday drives, when showing one's city to visitors or teaching one's children about their place. If, as Greenblatt (1991: 122) argues, 'everything in the European's dream of

FIGURE 30 Sightseers visiting Mount Coot-tha earlier this century had the spectacular views organised for them into a panoptic based on the compass. This type of early touring mimicked the surveying gaze of early colonialism. But these views were not for a distant metropolitan core, they were the beginnings of a settler touring practice that helped to constitute a sense of the self in the emergent nation home. (Source: Traill 1902)

possession rests on witnessing', then this type of tourism is the point at which 'witnessing' is orchestrated for the mass participation of the new locals. The promotion of Mount Coot-tha was intended to create a local tourist practice that not only bore witness to 'progress' but actually produced it by elaborating the settler's sense of belonging to and possessing the land (Morris 1988a). Through this local touring, the viewed fragment represented the emergent nation-home. The views actively produced and promoted at Mount Coot-tha guided tourists on their journey towards the destination of knowing and seeing a new homespace, of becoming Australian. The nation they saw contained a Nature tamed by colonisation and only traces of the Aboriginal inhabitants.

INDIGENOUS TOURINGS

In my reading of Brisbane's planned ecocentre I proposed that its mechanisms of incorporating Aboriginality relied on revitalised primitivist stereotypes which are fundamentally colonialist in effect. I was implying that 'predatory appropriation' – in that case, of a primitivised Aboriginality – is a persistent rule of cultural exchange in modern Australia (Morris 1988b: 267). This is a grim but familiar prospect which marks the ongoing efficacy of colonialist relations of power and suggests that the 'post' of postcolonialism is still distantly placed. But to categorise all cultural exchanges as inevitably or singularly constituted around appropriation has its own colonialist effects. Interpretations that are infatuated with appropriation are bound to uncover subordination as the positioning of the Other. They inevitably render passive those whose identities and cultural properties are seemingly appropriated, not only without consent, but without opposition, negotiation or, even, some unintentional consequence which might destabilise the colonialist foundations of the transaction. I am not proposing that I misread the Brisbane ecocentre. The scale and the level of commodification at the ecocentre may well relegate this otherness machine to no other fate than predatory appropriation. However, I do want to propose that in many emergent tourist ventures, big and small, private and public, a more complicated politics is present. Tourism ventures may well contain a familiar colonial politics but they may at the same time, contain hints of a more unsettling and, at times, optimistic postcoloniality.

What is precluded in the addiction to appropriation is not only the ontological presence of a possible postcolonialism, but more specifically the possibility of an anticolonialist criticism (Langton 1993: 7). Not least, an insistence on appropriation as the key dynamic of cultural exchange perpetuates a melancholic nostalgia for a passively fixed, but lost, object/subject. Žižek (1989b: 43) argues that in this nostalgic fascination, 'the gaze of the other is ... domesticated, gentrified'. Indeed he proposes that the very function of fascination is to 'blind us to the fact that the other is already gazing at us'. Interpretations that are fascinated with the ongoing appropriation of a primitivist stereotype in dominant discourse and cultural representation establish 'essential' identity – as I did in my reading of the ecocentre – as either present, but misused, or a fallacy of western desires. It denies the creative engagement 'primitivised' groups may have with such ascriptions and the implications of such engagements. It also denies the various ways in which such groups may themselves appropriate colonialist signifiers of power and centrality. As Langton (1993: 10) notes, Aboriginal cultural production increasingly 'consume[s] and reconsume[s] the primitive', just as it consumes and reconsumes other identifications associated with the centring of dominant groups. Aboriginality is in this sense 'a realm of intersubjectivity' which is 'remade over and over again in a process of dialogue' (ibid.: 33).

In the final part of this chapter I want to return to Mount Coot-tha and examine a recent example of the making of an Aboriginalised tourist space at one site on the mountain, J. C. Slaughter Falls. This example helps to articulate a more complicated politics of intersubjectivity associated with indigenous engagements with such industries of consumption. At its inception this project was thought of as a community arts, place-making project. But it also self-consciously opens out Mount Coot-tha to a new ecologically and culturally 'sensitive' practice of touring the country. This site traverses, then, a space which is somewhere between the large-scale, highly commodified venture of the ecocentre and the older, more embedded practices of touring as sightseeing.

J. C. Slaughter was a man of government, the Town Clerk and City Administrator of the Brisbane City Council in the middle years of the twentieth century. He had brought to Brisbane the basic structure of its modern local government adapted directly from the British models of Manchester and London. In Slaughter's view 'culture' was the icing on the cake of local government administration. For much of his career he oversaw the administration of the 'pedestrian and unimaginative' business of local government. But by 1964 he saw the possibility of a new direction for city administration which would foreground 'the cultural aspects of civic life' and 'improvement schemes for purely aesthetic purposes' (Slaughter 1964: 13). It was some thirty years after this that the real momentum of cultural planning took hold in Brisbane and, somewhat appropriately, saw the insertion of a 'cultural' site at a picnic spot named after J. C. Slaughter himself and located on the slopes of that familiar pleasure ground of Mount Coot-tha.

FIGURE 31 The painted images used at the J. C. Slaughter Falls walking trail were self-conscious copies of artworks to be found at pre-contact art sites throughout Queensland. The artists used only designs from areas to which they were connected by kin. As both artists had a complex lineage linked to a number of tribal groups they could claim associations with large tracts of Queensland. Additional ratification to use these images came by way of the Brisbane Aboriginal Council of Elders. (Source: Brisbane City Council with permission of the artists)

In mid-1993 artists Laurie Nilsen and Marshall Bell of Campfire Consultancy, a Brisbane-based Aboriginal visual arts company, completed a community arts project commissioned by the Brisbane City Council. The project was part of the Council's commitment to reconciliation in the International Year of Indigenous People. This gesture of reconciliation produced a mile-long walking trail at J. C. Slaughter Falls, Mount Coot-tha, some ten minutes' drive from the urban centre of Brisbane. The trail winds past a number of artworks including paintings, etchings and stencils on existing natural rock surfaces, tree carvings, stone arrangements and a ceremonial dance pit. The site has a iconography which has a complex hybridity constituted out of traditional significations of Aboriginal land knowledge as well as signifiers of colonial territorialisations. All the artworks are new but they have been executed in self-consciously 'traditional' style and draw on motifs which exist at 'pre-contact' Aboriginal sites to be found across the state of Queensland (Figures 31 and 32). Yet the conceptual template for the trail is based upon the creative appropriation of the map, that over-determined signifier of colonialism.

This community arts project was intended to restore Aboriginal cultural production and articulation to city space. It conforms with other 'cultural development' models proposed for the Brisbane region at this time. In a recent regional planning exercise a cultural development plan modelled itself on what it defined as 'indigenous practices'. In Aboriginal society, the plan argued, the arts and culture are 'integral' to rather than separate from 'the totality of economic, social and spiritual life' (Mercer and Grundy 1993: 33). This cultural development model actively worked away from the notion of 'exploitative, superficial' incorporations of indigenous cultures 'simply as objects to be viewed by "external" communities' (ibid.: 37). Indeed the plan warns that Aboriginal cultural property may not be available for general access or display according to 'western forms of museology and heritage classification' (ibid.: 34). Instead this cultural policy proposes a more fundamental adaptation of Aboriginal notions of attachment to place. In this model, an Aboriginal-style stewardship overlays European proprietorialism and is offered as a route for all Australians to feel a sense of belonging in their local place, their nation.

As a community arts project the J. C. Slaughter Falls art trail has done what many claim to do nowadays – it has created place out of space. This Aboriginalised place is most definitely for all to see, for 'public' consumption. A purpose-built pathway leads visitors from a car park and picnic area to the artworks, and signs erected along the trail provide interpretation of the meaning and content of the artworks. The trail offers the Mount Coot-tha

FIGURE 32 Artists Laurie Nilsen and Marshall Bell of Campfire Consultancy stand in front of the 'Main Gallery' at J. C. Slaughter Falls. This gallery contains a compilation of the individual motifs used at various points along the trail. A nearby interpretative sign explains that this is a 'map' of the trail and gives a lengthy description of the stories and meanings of the images. (Source: Brisbane City Council with permission of the artists)

visitor a new way to tour. The panoptic, extraterrestrial overview of earlier this century is replaced by a terrestrial close-up (Chambers 1987). Here imperial visions have given way to a more intimate 'indigenised' touring in which pleasure is taken from an embeddedness in the landscape and a planned exposure to the local. The 1990s tour of Mount Coot-tha is no less orchestrated than the ecotourism centre or the lookout, but these orchestrations are carefully 'hidden' by a contrived under-development.

In the Australian landscape the usual mode of creating Aboriginal places for public consumption is through cultural retrieval. Places of cultural significance to Aboriginal Australians are subject to archaeological excavation, scientific investigation and heritage designation. Oftimes it is precisely these processes that take previously 'unknown' places and set them in the 'public' art-culture system, a consumption economy (Clifford 1990: 141). Such places are not only heritage sites but, precisely because of this designation, also tourist sites. Once transformed into tourist sites, such places are subject to intensive management not simply to protect or conserve them, but also to ensure visitors receive a sanctioned understanding of their significance. The art trail of J. C. Slaughter Falls is not a cultural recovery of the heritage kind. The selection of Mount Coot-tha, and of J. C. Slaughter Falls in particular, for the new art site had more to do with a local settlement history of picnicking and walking, than with a specific Aboriginal interest which had been verified through scientific or heritage systems of classification.

The J. C. Slaughter Falls art trail is one of a number of recent place-making 'events' in Australia which have unsettled, if not usurped, the surety of colonial power. As a government-commissioned artwork, the J. C. Slaughter Falls trail has much in common with Michael Nelson Tjakamarra's commissioned mosaic for the forecourt of the new Federal Parliament House in Canberra. This project was variously read as appropriated primitivism and a symbolic indigenous reclamation of the heart of the nation (Kleinert 1988: 92–95; and, for response, Johnston 1988: 98–100). If the inclusion of Tjakamarra's mosaic in the design for Parliament House was indeed appropriation, then this appropriation was far from stable. As an act of opposition against the government's land rights policies, Tjakamarra later removed the central stone in the mosaic, allowing him to reclaim the symbolic status of the artwork and, in so doing, leave only an unauthorised original on the Parliament forecourt.

The use of traditional techniques and motifs in the J. C. Slaughter Falls project also resonates with the controversy surrounding a recent repainting of heritage-listed, pre-contact rock art sites by the Ngarinyin people. The Ngarinyin painted over one of the Wadjina art sites in their country with the authority of it being their right and responsibility to maintain the art sites.

But the Ngarinyin people's repainting of their artworks was interpreted by some as 'desecration' not only of the 'authentic object' but of what was now also claimed, through heritage listings, to be national property (Mowljarlai and Peck 1987; Michaels 1993). The very means by which the Ngarinyin performed their legitimate rights and responsibilities towards their land and culture was seen as illegitimate interference under the determinates of heritage investments in the authentic object. These are unruly place-making events which undermine colonial authority not by a shrill oppositional 'resistance' but by a subtle, subversive interplay with the colonial constructs of and jurisdictions over cultural property.

The political effects of such events must still be measured against the tenuous material dimensions of contemporary Aboriginal rights over land. Placement and displacement, in the material and legal sense of rights over land, do still matter as a measure of postcoloniality. None the less, the postcolonial potential of these sites is apparent in the way in which such place-making processes destabilise colonialist parameters of authority and authenticity around cultural production. Walter Benjamin (1992) notes, in the context of the art object, how the process of mass reproduction 'detatches the reproduced object from the domain of tradition' and argues that this then produces a problem of authenticity which is also a problem of authority. He goes on to suggest that this breach takes the art object away from the practice of ritual (which is the implied source of authority/authenticity) and into the practice of politics. In their ambiguous relationship to tradition these art events were indeed also political events. Benjamin has an ambivalent, at times pessimistic, take on this process of politicisation. But it is also possible to see this politicised reproduction as positive, as providing new ways for Aboriginal groups to articulate authority. These place-making projects take hold of colonialist constructs of people and place and rendered them unstable, no longer securely in the hold of colonialist arrangements of knowledge and power. They toy with the very adjudications of what are valid expressions of Aboriginality and the very hold of non-Aboriginal authority over such expressions.

The Aboriginal art trail at J. C. Slaughter Falls similarly unsettles notions of authenticity and authorisation. The 'traditional' designs used by the Aboriginal artists are not local Aboriginal designs but are based on motifs belonging to the Ngui, Mandandanji, Jiman and Kamilaroi peoples from areas well outside of Brisbane with whom the artists have various family ties. The project was executed and ratified by local Aborigines but was commissioned by the Brisbane City Council, at least in part, for a form of tourist practice. And it is in a place that was not previously of special significance to the indigenous producers. From within a position that has an investment in a primitivist original, here is the basis of a charge of inauthenticity; the production of an art object under an 'external' commission for public consumption (Cohen 1988: 376). But to designate this site as 'inauthentic' robs it of any postcolonial effect and seriously undermines the anticolonial intentions of the Aboriginal artists themselves.

In their use of traditional styles for the artworks, the artists of J. C. Slaughter Falls art trail might well be seen to be engaging in a form of strategic essentialism. The artists have reclaimed and redeployed the very motifs by which their culture has conventionally been taken up into non-Aboriginal models of art-culture consumption. That is, the artists have taken what is designated as authentically traditional and mobilised it in the pursuit of an anticolonial objective. Let me linger on this notion of strategic essentialism a little longer, for it is not only in the term 'essentialism' that trouble is encountered but also in the term 'strategic', which implies a cognisant intentionality. There is little doubt that Aboriginal identity politics, like other forms of identity politics, frequently adopts an 'operational' essentialism (Spivak 1990: 12). We can think of this, as Judith Butler (1990a: 325) does in the context of gender, as a provisional 'performative invocation of identity' within a less fixed reality of multi-constituted identities negotiating an uneven terrain of power and difference. Butler insists that this is not so much a process in which the instrumental subject 'wield[s] an essence at a difference' (1992: 109–110). Rather, it is a process in which a category 'by which the [subject] itself has entered into discourse' is remobilised. This mobilisation may or may not be linked to an instrumental intentionality. It may be a self-conscious mining of an externally ascribed identity or it may even be (and often is) precisely how an individual or group see themselves. But whatever the link to cognisant intentionality or to the blurred boundaries between external and self-ascription, the articulation will have effects, it will be strategic.

This is most clearly evident in the way in which the 'traditional copies' used in the artwork for this project generated a politically productive anxiety about authenticity. The shaky authenticity of these 'copies' was stabilised through recourse to the issue of authorisation, that is, how permission was gained to use these designs. Without such permission this art, regardless of its style, may well be seen by Aborigines as nothing more than an unauthorised image, perhaps even 'graffiti' (Michaels 1993: 61). The official description of the project made explicit the procedure by which the Brisbane City Council sought the permission of, and made a 'copyright' payment to, the Brisbane Aboriginal Council of Elders. This is a group of elders from various parts of the country who are now based in Brisbane. If one were to take a formal ethnographic view of the Council of Elders it is unlikely that they would necessarily have collective 'authority' either over the various traditional images used at J. C. Slaughter Falls or even the locality. But the presence or absence of this 'traditional' authority was not what was required by the City Council or needed by the artists themselves. The City Council's self-conscious documentation of the Elders' authorisation of the project was, in itself, part of a performance of authenticity and, grounded in 'tradition' or not, worked to give the site a more broadly registered Aboriginal legitimacy. Moreover, the artists' insistence on authorisation from the Elders established a protocol of consultation within the Brisbane City Council. In insisting on this procedure the artists sought not simply to ensure that their work at J. C. Slaughter Falls was seen to be legitimate, but also to establish a more general practice of Aboriginal consultation which reinstated indigenous authority over the city in a wider context (interview, Marshall Bell, October 1994).

Eric Michaels points out that 'traditionalism and authenticity' are completely false judgements to assign to contemporary Aboriginal practices of cultural production (Michaels 1993: 50). In this case the 'traditional origins' of the artwork have indeed a complicated trajectory which derives as much from the Aboriginal artists as it does from the desires of the commissioning body. The trail was a product of a relationship with a state-based commissioning body and it was incorporated, through tourism, into a 'global' art-culture system of consumption. It was produced within a theatre of Aboriginal authorisation, but the structure of this authorisation detoured from familiar models which invest authority in 'traditional' owners. But perhaps the most unsettling aspect of the trail is that the product does look just like a traditional Aboriginal art site. It is not just the impeccable copies of traditional motifs used in the art that capture this 'authenticity', but the very media itself: rock faces, trees and stones. The product reads as the real thing because it conforms with enduring colonial constructs which place Aboriginality in Nature.

REMAPPING THE COLONIAL

Meaghan Morris (1988a: 38), reiterating de Certeau, argues that tours (teleological narrative drives) and maps (fixations of the Present) are 'competing modalities in a process of narrative description'. This relationship between mapping and touring, fixing and moving, is clearly marked in the art trail of J. C. Slaughter Falls. This trail is not only based around 'copies' of traditional Aboriginal artwork, but is also an Aboriginal translation of that hallmark of imperial visioning, the map. Here, quite literally, the tour 'postulates maps' and the map 'condition[s] and presuppose[s]' the tour (ibid.). Much has been written of the power of the map within colonial expansionism (see, as examples, Harley 1988, 1992; Livingstone 1992). There is little doubt that the map is a 'form of mimetic representation' which 'textually represents the gaze through transparent space' (Blunt and Rose 1994: 8). Yet the emphasis on the hegemonic effect of the map may well overstate the power of the cartographic imagination. The encounters that led to the production of the ordered spatiality of the map and those that are precipitated under the guidance of the map always jeopardise cartographic certainties. This is not to suggest that the processes that challenge the map are not spatial or, indeed, precisely cartographic. Said (1990b: 77) argues that because the imperial project is an act of geographical violence the imagination of anti-imperialism is distinguishable by the primacy of the geographical. For the colonised, insurgency is in part a search for and restoration of place lost. At J. C. Slaughter Falls this restoration was realised not only by a return to tradition but also by a creative reappropriation of the map.

At the J. C. Slaughter Falls trail the artworks, and the stories they represent, retrieve Aboriginal spatio-cultural narratives of country. But these various Aboriginal representations of country are orchestrated into the spectacular 'whole' of the trail through recourse to the European cartographic imaginary. The trail begins with a schematic map, in an Aboriginalised style, of the entire trail. This map is both etched onto a rock and presented on a signpost located at the start of the trail. On the sign, the Aboriginal translation map is juxtaposed with a topographical map of the trail (Figure 33). The mapping concept is repeated in the artwork itself. The Main Gallery, for example, is a painted compilation, a schematic mapping, of the individual artworks along the trail (see Figure 32). The cartographic semiology of this

FIGURE 33 The trail at J. C. Slaughter Falls begins with a sign in which an Aboriginalised map and a topographic map of the path are juxtaposed. (Source: Brisbane City Council with permission of the artists)

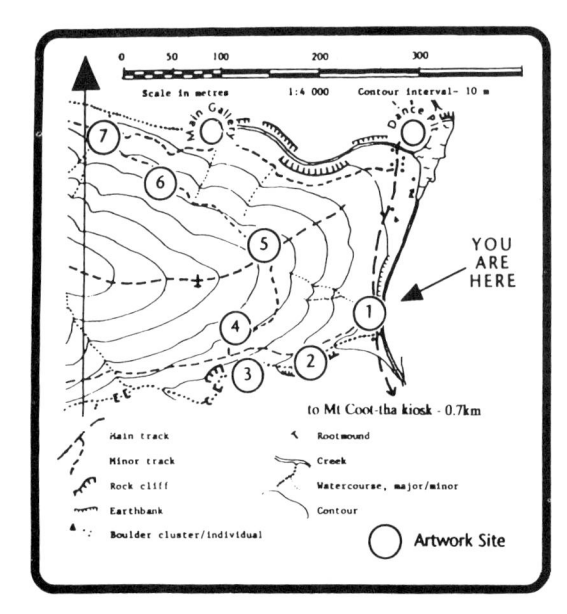

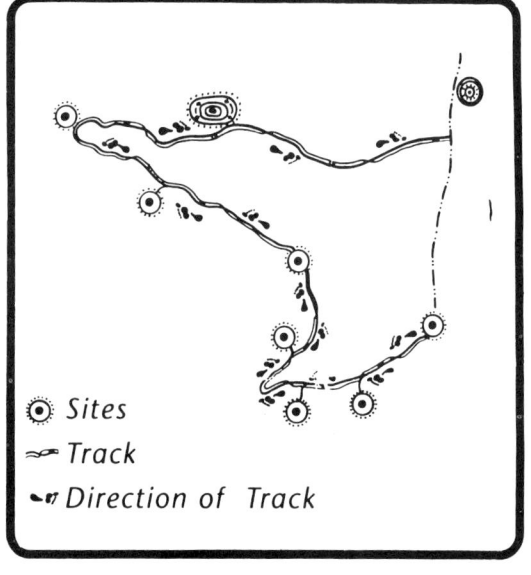

project parodies the semiology of colonialist territorialisations. Yet the mapping produced is a hybrid. The interpretative sign of the gallery provides an ordered spatial narrative to accompany this 'map':

> This main gallery is a map of the path you have just walked. The circles show each site and the footprints show which direction the track takes. The ochre stencils at site 2 represent the weapons used by the ancestral figure Biami to bring light into the world. The figure Biami (site 3) was either etched or painted on a rock surface depending on the texture or hardness of the surface. Biami figures can be found in Queensland's Carnarvon Ranges.
>
> Rocks were used for various purposes: cooking, fish traps, ceremonies, stone arrangements and artifacts such as grinding dishes and grinding stones (nardoos). The stone arrangement at site 4 depicts Munta-gutta the creator spirit. Site 5 depicted in this painting shows a tree carving from the Kamilaroi nation known as Gobi, the two headed goanna which is friendly to everyone. Site 6 shows Munta-gutta's eggs in the river which it guards from intruders. Site 7 shows a set of footprints of a kangaroo standing still and then on the move with each hop getting further apart as it picks up speed.
> (J. C. Slaughter Falls art trail, Main Gallery, interpretative sign, 1993)

An Aboriginalised cartographic map of Brisbane city was also designed by Laurie Nilsen, and is to be used not only as the logo for this site but as a city-wide signifier for places and sites with Aboriginal content and meaning (Figure 34). Like the trail, the logo is a gesture which symbolically repossesses colonised space through a cartographic imagination in translation.

In repossessing the map and placing it alongside reactivated indigenous mappings, the J. C. Slaughter Falls art trail has a special power which derives from a 'doubleness', its 'unsteady location simultaneously inside and outside the conventions, assumptions, and aesthetic rules which distinguish and periodise modernity' (Gilroy 1993: 73). The inventive renegotiation of the map leaves it transformed, somewhat depleted, of its colonialist powers. In actively engaging with the language of cartography there is an escape, without a leaving, from the inevitability of the perspectival positioning of colonialism (de Certeau 1984: xiii). The semiology of the J. C. Slaughter Falls trail has an enticing doubleness which destabilises the boundaries between Self and Other, colonial and traditional, authentic and inauthentic. This hybridity unsettles the comprehensive hold of colonialist constructs.

The postcolonial power of this counter-cartography is also tied to the

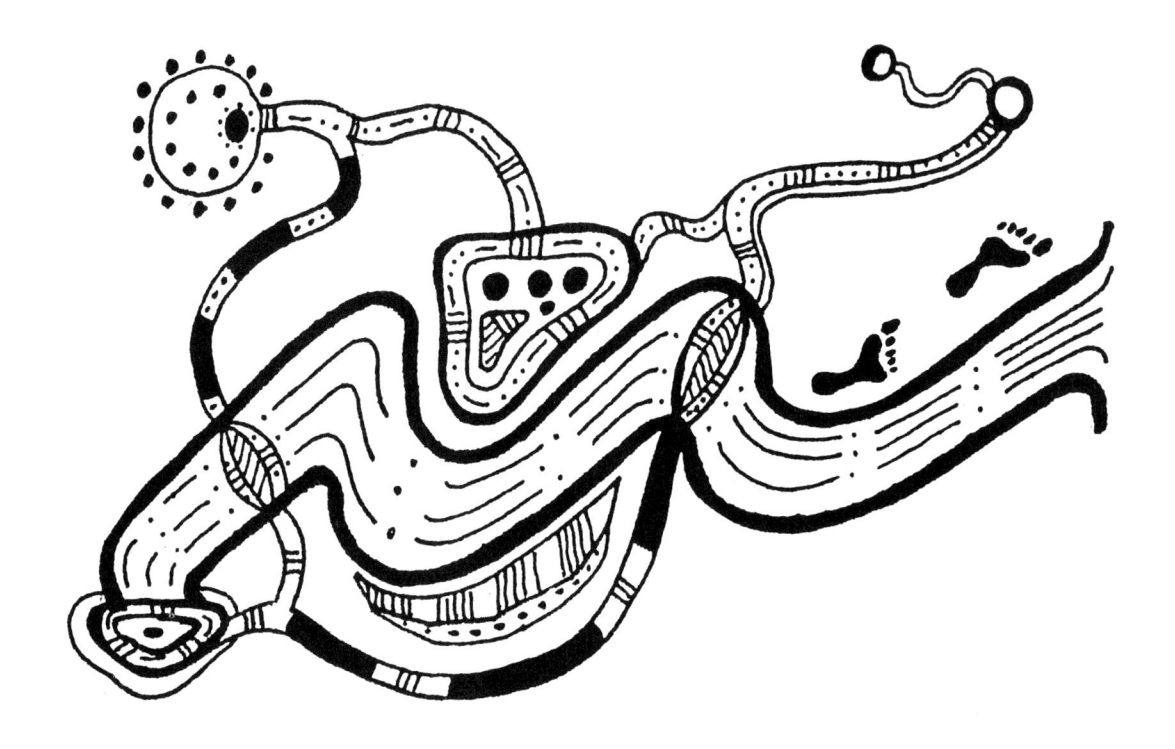

FIGURE 34 Laurie Nilsen's Aboriginalised remapping of the city of Brisbane. This logo is to be used by the Brisbane City Council to mark all Aboriginal community arts, place-making projects and all Aboriginal heritage trails. The logo shows the main river, the city centre and major roads and bridges. (Source: Brisbane City Council with permission of the artists)

actual processes of production and the subsequent use of the art trail. J. C. Slaughter Falls was chosen as a location for an art trail not because of its importance to the local Aboriginal community, which itself is more likely to trace connections to places outside of Brisbane. The presence of the art and the dance pit may ensure J. C. Slaughter Falls becomes a special place of significance for Brisbane Aborigines, a place for gatherings, school excursions, Sunday picnics. But the trail is not intended to be exclusively for Aboriginal use. In its original conception the trail was designed for the pleasure and education of general visitors to the area. The J. C. Slaughter Falls art trail is a self-conscious, bicultural remapping of a small segment of urban open space intended to teach those who walk the trail another way of seeing the land. The official documentation for the project explains that the pedagogical function of the site is intended to create 'a better understanding and appreciation of Aboriginal culture' (Brisbane City Council 1993b: 10). The J. C. Slaughter Falls trail has a directed hybridity. It is meant for those unknowing people, settler Australians, those people the Aboriginal writer Paddy Roe (1988) calls the 'children' of the country. The art trail presumes, then, not a colonial authority over country but an Aboriginal authority. It presumes that Aboriginal knowledge needs to be taught, needs to be learnt. This is not simply another version of non-Aboriginal appropriations of indigenous knowledge, although clearly the commissioning City Council was amenable to the idea; it is also an indigenous engagement with such western desires.

Apart from general visitors, one of the anticipated users of the trail was a local Brisbane orienteering group. This group was consulted by the artists and provided the topographical maps used in the interpretative material. Few pursuits are so dependent on cartographic knowledge as orienteering. Orienteering is a practice of finding; it re-enacts the process of discovery and spatial ordering associated with the pragmatics of making empires. Each orienteering exercise relives the thrall and challenge of discovery, the skill of finding pathways through space, locating markers. It is the play of making spatial sense of the 'unknown', where the ordering of space around oneself, the methodical finding and registering of visits to places, aided by map and compass, leads to 'home' and the triumphant end-point of the adventure. The Aboriginal remapping of J. C. Slaughter Falls attempted to introduce a new logic, an Aboriginal logic, to this play of movement through country. But in its emphatic pedagogy of Aboriginal meaning, in its single one-way path, in its mapped and remapped traditions, this 'uncanny' path was too fixed, too predictable for the orienteerer. It seems that the path is frequently by-passed by orienteerers or only partially incorporated into their activities: it has become a single point on a map not of Aboriginal making.

The pedagogical purpose of the site was also extended to younger Aboriginal men and women. From the outset the artists determined that the production of the original trail, and any subsequent trails, would serve the purpose of educating younger Aborigines in the techniques of traditional-style art production as well as the ways of caring for the country. The production of the trails became a focus for art practice in which the more experienced artists trained less experienced ones and established the foundations for an ongoing cultural association with these sites. Artist Marshall Bell's longer term vision for the sites sees them incorporated into a community practice of site care and maintenance which requires that the sites be regularly repainted and reworked. Indeed the materials used for painting (ochre and glue) were chosen not just because they were traditional but because they had a short life and the art sites would need to be 'restored, looked after, as a continuation of our culture' (interview, Marshall Bell, October 1994). An intended ephemerality in the artwork becomes the mechanism for the re-emergence of a durable cultural practice. In this sense, the anticolonial 'tradition' of the art trails extends beyond the specific art object and into a radical reactivation of Aboriginal cultural production in the urban scene.

Said (1990b: 78–79), talking about insurgent nationalisms and their use of place-based myth-making, argues that such creative processes of reclamation present a 'new territory'. This is a space that is neither the panoptic, flattening space of imperialist visions nor a restoration of the pristine or pre-modern. The J. C. Slaughter Falls art trail is a remapping, which is intended to herald a new territory. Its hybrid form suggests the demise of the persistent and static binary oppositions that are so fundamental to the culture of colonialism. But the J. C. Slaughter Falls art trail was not produced in order to liquidate the distinction between Self and Other, but to reorder this difference in a way that empowers and liberates Aborigines. Here is a site that is precisely, literally, a 'third space' constituted out of the power-laden, 'dialectical play of "recognition"' between Self and Other, colonist and colonised (Bhabha 1985: 156). Within this space, 'the meanings of colonial inheritance', which include primitivist stereotypes as much as colonialist cartographies, were transformed into the beginnings of 'liberatory signs' (Bhabha 1994: 38). This to accept that *in some forms* the re-Aboriginalisation of place, although not land rights in itself, can be a meaningful reterritorialisation. Indeed it is precisely because this type of reclamation occurs outside of the debilitating and primitivist proofs of traditional lineage and land association still required by heritage and land rights jurisprudence that it offers a most democratic possibility for those groups wishing to remake their mark over land. The trail has a productively unfixable 'double heritage' (Gates 1984: 4) which may in fact lead to a place where we need to be.

This is not to propose a joyous transcendence of colonialism by way of ethical industries of consumption. This art trail probably will not result in Aborigines gaining significant or meaningful land rights in relation to Brisbane. Indeed, while the map of the colonisers exercised its conceptual power through the force of contact, this remapping may remain symbolic. As recently unsuccessful Mabo-style claims to Brisbane testify, it is unlikely that this city will be repossessed, that *this* map will, as Baudrillard (1983: 2) suggests is possible, precede a fully reclaimed Aboriginal territory. The Aboriginal art trail at J. C. Slaughter Falls may not usurp the comprehensive material hold that neo-colonialist and imperialist activities have on the city of Brisbane. In providing yet another mechanism for (a now indigenised) touring, it may even permanently defer this very possibility. It will not preclude Aboriginal culture from being 'appropriated' in this or other settings. It will not preclude the possibility, already suggested by the site's (mis)use by orienteerers, of what Walter Benjamin (1992: 232) refers to as 'consumption . . . in a state of distraction'. But despite these material limits, it is one space that does begin the necessary task of reminding non-Aboriginal Australians, albeit temporarily, that they are strangers in their own land.

It is undeniable that a persistent dimension of the consumption-based tourism industries is their colonialist tendencies. In different ways, and to greater and lesser degrees, the tourist sites visited in this chapter are colonialist. At times this is evident through the mastery of Nature and the imaginative obliteration of Aborigines, at times through the commodified or primitivist conflations of Nature and Aboriginality. That these processes persist, and virulently do so in certain spheres of modern life, is unarguable. Nevertheless, this chapter has also shown that colonialism is a varied and inherently unstable project. Not only is tourism one of a number of forms that colonialism might take in the modern world, but even within tourism industries themselves, colonialist tendencies vary according to historical and material specificities.

It is commonplace for colonialism in the present to be read primarily as a process of cultural 'appropriation' in which colonised cultures are commodified and contaminated. The massive critique that emerged during the 1980s of tourism and heritage industries was in part driven by an anxiety that such industries commodified authentic objects, settings and cultures and in so doing wrenched them away from their essential point of origin and rendered them fake. This argument is often elaborated by suggesting that commodification entails an invasion of the local by 'alien', globally linked, processes which break the formative chain between an unamended local and the emergence of autonomous (read authentic) cultural formations. Many of the critical accounts of heritage and tourism are underpinned by the assumption that these industries are a danger to the survival, let alone the articulation, of more real places and social identities.

Appropriation is by no means the universal rule of the cultural exchanges that occur within the art/culture/nature tourism industries. It may well be that, rather than tourism and heritage industries being a threat to 'authentic' places or identities, they produce the melancholic fantasy that there are (somewhere) or were (sometime) authentic places and peoples. If the binary of an unmediated authentic and a mediated inauthentic is a nostalgic by-product of modernity itself, it is hardly an adequate framework for understanding the complex, intersubjective politics of place. A critical perspective that harbours a nostalgia for the authentic does not simply miss the point of certain contemporary forms of cultural production that toy with the past, it also robs them of their political power in the present.

The constitution of identity by colonised groups inevitably entails a complex interplay of the past and the present, self ascriptions and social designations, intentioned action and unpredicted surpluses. These articulations of identity may even be made, and often are made, in terms that not only claim but also rely on notions of essential origins and authenticity. In this sense they can entail an active engagement with 'alterity' (Appiah 1991: 354). But such articulations are necessarily made under the conditions of modernity; within, and often against, the specific opportunities and restraints produced by a history of colonisation, including various systems of legal adjudication, regimes of signification and processes of commodification. For colonised groups, the reappropriation and redeployment of primitivised Otherness are not simply about returning to the pre-modern but precisely about engaging in a modern politics of identity and place.

7
CONCLUSION

•

In 1988, the year in which Australia 'celebrated' the bicentenary of its founding as a settler colony, an Aboriginal activist visited Britain. On a windy day on Brighton beach, surrounded by invited journalists, Burnam Burnam raised the Aboriginal flag and declared the British Isles to be Aboriginal territory. This colonial return mimicked Governor Phillip's hoisting of the Union Jack at Sydney Cove on the east coast of Australia 200 years before. Both were symbolic events. The first marked the 'beginning' of the British territorialisation of the continent that came to be known as Australia. The second successfully parodied the audacious banality of that event. In 1788 those few Aborigines who witnessed the raising of the Union Jack could not imagine what would happen – to them, to all Aborigines – as a result. In 1988 the millions of Britons to see the media reports of the Aboriginal flag being hoisted on Brighton beach could assume it did not mark the 'beginning' of anything much at all. One was an inaugural event which, through the force of desire and sheer might, opened out into a history of colonisation. The other was a memorial event which, despite the force of desire, could neither claim Britain nor undo Australia's history of colonisation. The embedded unevenness of power, which is the legacy of imperialism, meant that these events did their symbolic work in quite different ways. Like these events this book has brought Australia and Britain back into 'contact'. But this is not a cause-and-effect encounter. It is, like many 'first contacts', one in which the two sides see each other, impinge upon each other, but do not recognise the full weight of the histories and geographies implied by their meeting.

GEOGRAPHICAL ENCOUNTERS

This book has sought to create a productive encounter between new theorisations of imperialism and postcolonialism and the specific space of the contemporary city. I have undertaken this task in order to give geographical expression to a theoretical field that is rich in its allusions to space but often poor in its elaboration of the real worlds in which this spatiality operates. The location of much of the current theory within the fields of literary criticism and history has meant that its relevance to the conditions of everyday life in the present is often oblique. And while much of this theory is about difference, about deconstructing master narratives, about space, these concerns are often expressed through grand theory and not through the fundamentally deconstructive space of the local. I am not simply suggesting that the local provides exceptions to the rules, although this is often the case. I am proposing that through attending to the local, by taking the local seriously, it is possible to see how the grand ideas of empire become unstable technologies of power which reach across time and space.

The generic 'local' that has formed the focus of this book is the contemporary First World city. This focus has not been in pursuit of a new model of global urbanisation or to embellish the nature of the 'postmodern' urban. Indeed, these urban studies self-consciously work away from both possibilities. Old models of urban development which placed the colonial city as a mid-point in an evolution from pre-modern to modern have outlived their usefulness. It is not that the distinction between core and periphery, haves and have nots, has gone away – it is devastatingly present. But the 'where' of this geography is increasingly confused: First World cities have their Third World neighbourhoods, global cities have their parochial underbellies, colonial cities have their postcolonial fantasies. Urban transformations such as gentrification, consumption spectacles and heritage developments, are regularly understood as postmodern. But these spectacles of postmodernity are entwined in a politics of race and nation which cannot be thought of constructively without recourse to the imperial inheritances and postcolonial imperatives that inhabit the present.

The 'real worlds' of this book are of course not simply material worlds. Imaginary and material geographies are not incommensurate, nor is one simply the product, a disempowered surplus, of the other. They are complexly intertwined and mutually constitutive. Together they gave energy and drive to the territorialisations that constitute imperialism. Together they have created the most painfully uneven geographies of advantage and disadvantage. The social construction of space is part of the very machinery of imperialism. In the name of the imperial project, space is evaluated and overlain with desire: creating homely landscapes out of 'alien' territories, drawing distant lands into the maps of empire, establishing ordered grids of occupation. These spatial events did not simply supplement the economic drive of imperialism, they made it make sense; they took it from the visioned to the embodied, from the global reach of desire to the local technologies of occupation. They established the beginnings of that most permanent legacy of imperialism: the contest between that which, through space itself, has been 'naturalised' and that which has been made 'illegitimate'.

UNRULY IMPERIALISM

Imperialism has neither a uniform realisation nor a static persistence. Imperialist constructs have taken an unruly passage across time and space. The hold of imperialist regimes of power is tied to the very uncertainty they face in their manifestation on the ground: in their encounter with the unpredictability of the Other and the inconsistency of the Self. In the face of this uncertainty imperialism must always reinscribe its frames of power and difference and this is what helps to give it its tenacity. Space is a crucial component of this anxious articulation of imperial authority.

The ordered spatialities of here and there, Self and Other, which were imagined in the inaugural moments of the imperial project, were regularly unsettled by the disorderly encounters forged during colonisation. In Australia the imperial gaze imagined a land unoccupied but encountered a land most surely peopled. Colonial constructs of space, such as the ordered plans for the city, struggled to contain the subversion created merely by Aborigines being present. That some 200 years later the Waugal Dreaming can 'appear' in Perth and unsettle contemporary urban redevelopment plans points to the failure of such technologies of containment to realise comprehensively the imaginary construct of *terra nullius*. In British cities, too, imperialism meant change and in particular new levels of industrialisation and urbanisation. While there was pride in this imperial growth, it also spawned anti-urban organisations which are still present over a century later, and which through their conservation efforts regularly activate the memory of empire.

In the contemporary moment the always contingent order of imperial geographies has been further undone. Formal decolonisation and postwar migration and settlement have brought an embodied edge of empire into the heart, while the demise of empire has meant that Britain now has different global and regional affiliations. These changes have not marked the end of empire – a pure postcolonialism – but established the conditions for revised imperial articulations. Imperialism lingers in the present as the idea of empire itself, as a trace which is memorialised, celebrated, mourned and despised. This is a potent memory which can shape trajectories of progress, drive nostalgic returns and establish the structures of difference through which racialised struggles over territory operate. In the City of London, for example, imperial nostalgias are not simply present as a residual past. They are sanctioned and activated by conservation practices which influence, in most material ways, the very course the City takes into the future. Here as in the various other case studies, tradition is not simply about an escape from modernity but about negotiating modernity, about being modern. Indeed, it is precisely the desire to memorialise empire that has helped to drive the City beyond its traditional boundaries and into areas like Spitalfields where Bengali Britons are now struggling to make a home-space in a new nation.

Imperialism also comes into the present through 'new' forms of exploitation and domination, what might be thought of as neo-colonialism. Nineteenth-century imperialism has given way to new regimes of desire. Otherness is no longer a repressed negativity in the constitution of the Self, but a required positivity which brings the Self closer to, say, a multicultural present or an ecological future. Contemporary spaces of consumption seek out Otherness. But it is not just any Otherness that is required. In tourism developments in Australia it is an essentialised Aboriginality, honed for the spectacle, which is taken up into these systems of commodification. In these new tourist spectacles Aboriginality is, once again, returned to Nature. Early imperial constructions of Aboriginality and Nature confirmed the division between an 'uncivilised Native' and a 'civilised Culture'. In the current unison, Aboriginality serves as a template for the 'proper' return of Civilisation to Nature, the source of an ecological Self. But these eco-desires also serve as a means for settler Australians to develop their own sense of being in the land and, as such, of a more final and complete colonisation. Of course, such returns are still likely to be regulated by the requirements of capital. In the city of Perth, for example, the proposition to return development land to Nature was seen as 'unproductive' and deeply subversive to city development aspirations.

In Spitalfields too the desire for Otherness became part of the politics of place. Gentrifiers and developers regularly celebrated the distinctive 'multi-cultural' history of the area. But this celebrated cohabitation could not be too promiscuous or unpredictable. Here a multiculturalism of convenience emerged based on a properly placed (spatially segregated) Bengali community. Ordered and domesticated, the Bengali residents of Spitalfields could become a safe, present-day supplement to the narrative construction of Spitalfields as the emblematic place of an embracing, tolerant Englishness.

In the mapping of current imperialisms, appropriation is often posited as a key dynamic in the exercise of power. A prime example of this is the way in which tourism and heritage industries reproduce essentialised constructions of identity and take these up into systems of commodification. It is as if the stark territorial appropriations of nineteenth-century imperialism have given way to more fractured patterns of cultural appropriation. As I have shown, such appropriations do occur and they can have imperialist effects. Essentialist constructions of Otherness do work to categorise colonised groups within desired and confining templates. Nevertheless, it is a form of imperial nostalgia, a desire for the 'untouched Native', which presumes that such encounters only ever mark yet another phase of imperialism. Colonised groups do not enter passively into such systems of commodification, nor are they ever neatly stitched into place within them. Essentialised constructs of identity and place are open to a range of reinventions, adaptations, invigorations and reappropriations at the hands of both colonisers and colonised. The encounters that occur through new processes of commodification might well help colonised and diasporic groups to articulate a sense of self in productive new ways: giving local political struggles a global reach; providing new arenas for the elaboration of tradition; opening up new economic opportunities; and establishing influential systems of pedagogy. Commodification is not simply a process by which the colonised, the 'native', 'tradition' is corrupted.

The idea for Banglatown in London shows precisely how problematic it is to locate imperialism in the processes of appropriation/commodification and the construction of essentialised subjectivities. Banglatown was not imposed on the Bengali businessmen, nor did they produce this idea unaware of their strategic toyings with the category 'Bengali'. Here the Bengali businessmen actively engaged in elaborating essentialist constructions of identity through commodified systems in an attempt to wrest control of power and space. Similarly, the Aboriginal artists who entered into the 'reconciliation' place-making project in Brisbane inventively redeployed 'tradition' in order to establish new protocols of influence over the modern city. In both cases, different though they may be, the notion of sure-footed imperialist appropriation is fundamentally unsettled. Also in these cases, essentialised constructions, although reaching back into the past, are produced in the present in order to negotiate the inequities of power produced in the modern.

POSTCOLONIAL POSSIBILITIES

By problematising the dynamics of appropriation and essentialism I do not simply strive towards an emergent postcolonialism by way of the Native. The tenacity of an adaptive imperialism is a reminder of the fantastic optimism of the term 'postcolonialism'. While indigenous land rights claims and postwar settlements in Britain might be thought of as postcolonial formations, their truly *post*colonial effect is still faintly traced. It is indeed hard to imagine a moment that is beyond imperialism. In this sense the postcolonial is not so much about being beyond colonialism as about attending to the social and political processes that struggle against and work to unsettle the architecture of domination established through imperialism. This includes the 'space clearing' gestures of intentional anti-imperial politics. But it also includes the various uncontained excesses and unsettlingly hybrid outcomes of the cohabitation produced by imperialism itself. For example, the anticolonial intention of the Aboriginal claims over Perth were amplified by the anxious reactions of non-Aboriginal Australians to the 'appearance' of the Waugal Dreaming in the secularised city. Similarly, the Banglatown plans for Spitalfields actively sought to create alliances with 'big' capital in pursuit of creating a Bengali place in Britain. This created a crisis of affiliation between the Bengali community and the very groups who saw themselves as their guardians and their anti-racist allies.

Within the current discussion of the postcolonial much emphasis is given to the disruptive power of hybridity. In placing hybridity as a key signifier of postcoloniality the inevitable vulnerability of colonial structures of power and categorisation is properly exposed. Imperialism undoes itself, against itself. Nevertheless, in thinking through the postcoloniality of the hybrid form it is often too easy for agency and intentionality to be displaced: that is, for the postcolonial to be posited merely as the 'subject-effect' of the colonial. In the example of the Brisbane 'reconciliation' place-making project a most decidedly and self-consciously hybrid place was created out of a creative redeployment of 'tradition' and a destabilising appropriation of the colonial construct of the map. Understanding the postcolonial effects of this project required more than relishing its visual hybridity. Here hybridity was not about the dissolution of difference, or even only about destabilising the constructs of Self and Other, but about renegotiating the structures of power built on difference. The political weight of this hybrid space resides not simply in its surface form and in its effects, but also in the intentional politics of its production, its desire to reassert Aboriginal authority over the space of the city. Attending to such intention, even while it might fracture into a disorderly effect, is a necessary component of a postcolonial critique.

Throughout this account of the politics of identity and place there have been various versions of what might be thought of as a return to origins – the claim that identity is 'given' through some uncontested inheritance or static place-based genesis. While the fractured and contingent nature of identity is undeniable, so too is the necessity of temporary fixings of identity around such essentialised notions. Such claims are deployed by the powerful to legitimate their rights over territory, to categorise Otherness as 'outside' and to domesticate difference. That is, claims of origin can be hegemonic. But claims of origin, such as strategies of fixing identity in place, are also important for marginalised groups who want to distinguish their claims from the hegemonic. Proposing that essentialist notions of identity and place are social constructs, and strategic ones at that, destabilises a whole range of claims for rights over space which are argued through the idea of origin. It is one thing for such a generalisation to unsettle nostalgias for empire or the violence of some nationalisms; it is another thing when it also compromises the claims for land made by colonised groups who are still intensely marginalised. That is, theoretical generalisations about the socially constructed nature of essentialised identity have an uneven political consequence which is far from incidental to ongoing political struggles.

In a contemporary world, constituted out of complex processes of deterritorialisation and reterritorialisation, movement and cohabitation, it may well be that what Kristeva (1993: 2) calls the 'cult of origins' needs to give way to a sense of place which is built around fractured vectors of connection and histories of disconnection. There is little doubt that the present has seen a radical 'unbuilding' of the geographies of imperialism, but it is also true that these new geographies do not surpass their past. They are made out of the continued negotiation of lingering and newly formed imperialisms as much as they are made out of the hope of postcolonial futures. It is because of the undeniable persistence of domination that claims of origin, the strategic formation of fixed identity, continue to be part of the politics of the present. That geography which Doreen Massey (1993a) refers to as a 'progressive sense of place' could surely only flourish within a more even, more democratic, terrain of power than that which characterises the present.

If postcolonialism is always haunted by colonialism, can there be a postcolonial geography? Can the spatial discipline of geography move from its positioning of colonial complicity towards producing postcolonial spatial narratives? The postcolonial geographies traced in these pages have not sought out a 'pure' postcoloniality. Instead, they have concerned themselves with the unruly fortunes of colonial constructs as they fold in on themselves, are recharged by new contexts or fundamentally challenged by their encounter with the colonised. They have not simply sought out Otherness in order to elaborate the nature of the imperial core or the inappropriateness of its colonial projects. Rather, they have engaged with anticolonial political imperatives. These postcolonial geographies have replaced the security of the maps of the past with the uncertainty of touring the unsettled spatialities of power and identity in the present.

BIBLIOGRAPHY

Abu-Lughod, J. (1980) *Rabat: Urban Apartheid in Morocco*, Princeton, NJ: Princeton University Press.

—— (1984) 'Culture, "modes of production" and the changing nature of cities in the Arab world', in J. Agnew, J. Mercer and D. Sopher (eds) *The City in Cultural Context*, London: Allen and Unwin, 44–117.

Ackroyd, P. (1985) *Hawksmoor*, London: Abacus.

Adam, I. and Tiffin, H. (eds) (1991) *Past the Last Post: Theorizing Post-colonialism and Post-modernism*, Hemel Hempstead: Harvester-Wheatsheaf.

Age, The (22 February 1989) 'Waugal reawakened'.

Agnew, J. and Corbridge, S. (1995) *Mastering Space: Hegemony, Territory and International Political Economy*, London and New York: Routledge.

Ahmad, A. (1992) *In Theory: Classes, Nations, Literatures,* London and New York: Verso.

Alexander, I. (1992) 'City centre planning for public or private interest', in D. Hedgecock and O. Yiftachel (eds) *Urban and Regional Planning in Western Australia: Historical and Critical Perspectives*, Curtin University of Technology, Perth: Paradigm Press.

Alsayyad, N. (1992) 'Urbanism and the dominance equation', in N. Alsayyad (ed.) *Forms of Dominance: On the Architecture and Urbanism of the Colonial Enterprise*, Aldershot: Avebury, 1–26.

Amery, C. and Cruickshank, D. (1975) *The Rape of Britain*, London: Elek.

Ancestors of the Swan River People (1989) *1833–1989 = 156 Years of ?*. Publicity broadsheet. Perth: Ancestors of the Swan River People.

Anderson, K. (1989) 'Cultural hegemony and the race-definition process in Chinatown, Vancouver: 1880–1980', *Environment and Planning D: Society and Space* 6, 2: 127–151.

—— (1990) '"Chinatown re-oriented": a critical analysis of recent redevelopment schemes in a Melbourne and Sydney enclave', *Australian Geographical Studies* 28, 2: 137–154.

—— (1991) *Vancouver's Chinatown*, Montreal and Buffalo: McGill-Queen's University Press.

Anderson, R. (1988) 'Meaning in the Urban Environment', unpublished Ph.D., Oxford Polytechnic: Centre for Urban Design.

Ansara, M. (1989) *Always Was, Always Will Be*, Sydney: Jequerity Pty Ltd.

Appadurai, A. (1990) 'Disjuncture and difference in the global cultural economy', *Public Culture* 2, 2: 1–32.

—— and Breckenridge, C. A. (1992) 'Museums are good to think: heritage on view in India', in I. Karp, C. Mullen Kreamer and S. D. Lavine (eds) *Museums and Communities: The Politics of Public Culture*, Washington and London: Smithsonian Institution Press, 34–55.

Appiah, K. A. (1991) 'Is the post- in postmodern the post- in postcolonial?', *Critical Inquiry* 17: 336–357.

—— (1992) *In My Father's House: Africa in the Philosophy of Culture*. New York and Oxford: Oxford University Press.

Architects' Journal (25 July 1990) 'RFAC Scorns Spitalfields Plan', 11.

—— (22 September 1993) 'Joint venture boosts No. 1 Poultry plan', 10.

Artley, A. and Robinson, J. M. (1985) *The New Georgian Handbook: A First Look at the Conservation Way of Life*, London: Ebury Press, Harpers and Queen Publications.

Ashcroft, B., Griffiths, G. and Tiffin, H. (1989) *The Empire Writes Back: Theory and Practice in Post-colonial Literatures*, London: Routledge.

—— (1995) 'Introduction: issues and debates', in B. Ashcroft, G. Griffiths and H. Tiffin (eds) *The Post-colonial Studies Reader*, London and New York: Routledge, 7–11.

Asian Herald (23–30 November 1990) 'Banglatown Ghetto'.

Australia Council (1990) *Arts Facts: International Visitors and Aboriginal Arts*, Research Paper No. 4, Policy and Research, Strategic Development Unit, Canberra: Australia Council.

Australian, The (19 June 1989) 'Perth covers 1000 sacred sites'.

—— (23 July 1989) 'Swan Brewery continues'.

Bailey, N. (1990) 'Community development trusts – a radical third way?', in J. Montgomery and A. Thornley (eds) *Radical Planning Initiatives: New Directions for Urban Planning in the 1990s*, London: Gower, 150–164.

Baker Harris Saunders (1988) *No.1 Poultry Public Inquiry: Proof of Evidence*, London: Department of Environment.

Baker, L. M. and Mutitjulu Community (1992) 'Comparing two views of the landscape: Aboriginal traditional ecological knowledge and modern scientific knowledge', *Rangeland Journal* 14, 2: 174–189.

Bal, M. (1991) 'The politics of citation', *diacritics* 21: 25–45.

Balfour, A. (1990) *Berlin: The Politics of Order, 1737–1989*, New York: Rizzoli.

Barnes, T. and Duncan, J. (eds) (1992) *Writing Worlds: Discourse, Text, and Metaphor in the Representation of Landscape*, London and New York: Routledge.

Barthes, R. (1981) 'Semiology and the urban', in M. Gottdiener and A. Ph. Lagopoulos (eds) *The City and the Sign: An Introduction to Urban Semiotics*, New York: Columbia University Press, 87–98.

Baudrillard, J. (1983) *Simulations*, trans. P. Foss, P. Patton and P. Beitchman, New York: Semiotext(e).

Bauman, Z. (1992) *Imitations of Postmodernity*, London and New York: Routledge.

Benjamin, W. (1992) *Illuminations*, London: Fontana.

Berman, M. (1982) *All That is Solid Melts into Air: The Experience of Modernity*, London: Verso.

Bernard Williams and Associates (1986) *Spitalfields Market Redevelopment: Economic Appraisal*, London: BWA report for London Borough of Tower Hamlets.

Bhabha, H. (1985) 'Signs taken for wonders: questions of ambivalence and authority under a tree outside Delhi, May 1817', *Critical Inquiry* 12, Autumn: 144–165.

—— (1990a) 'The other question: difference, discrimination and the discourse of colonialism', in R. Ferguson, M. Gever, T. T. Minh-ha and C. West (eds) *Out There: Marginalization and Contemporary Cultures*, New York: Museum of Modern Art/Cambridge, MA and London: MIT Press, 71–88.

—— (ed.) (1990b) *Nation and Narration*, London and New York: Routledge.

—— (1992) 'Postcolonial authority and postmodern guilt', in L. Grossberg, C. Nelson and P. Treichler (eds) *Cultural Studies*, London and New York: Routledge, 56–68.

—— (1994) *The Location of Culture*, London and New York: Routledge.

Binney, M. (1984) *Our Vanishing Heritage*, London: Arlington Books.

—— and Lowenthal, D. (eds) (1981) *Our Past Before Us: Why Do We Save It?*, London: Temple Smith.

Birch, T. (1992) '"Nothing has changed": the making and unmaking of Koori culture', *Meanjin* 51, 2: 229–246.

Bird, J. (1993) 'Dystopia on the Thames', in J. Bird, B. Curtis, T. Putnam, G. Robertson and L. Tickner (eds) *Mapping the Futures: Local Cultures, Global Change*, London and New York: Routledge, 120–135.

Blain, D. (1989) 'A brief and very personal history of the Spitalfields Trust', in M. Girouard, D. Cruickshank and R. Samuel (eds) *The Saving of Spitalfields,* London: The Spitalfields Historic Buildings Trust, 1–34.

Blaut, J. M. (1987) *The National Question: Decolonizing the Theory of Nationalism*, London: Zed Books.

—— (1992a) *1492: The Debate on Colonialism, Eurocentrism and History*, Trenton, NJ: Africa World Press.

—— (1992b) 'The theory of cultural racism', *Antipode* 24: 289–299.

—— (1993) *The Colonizer's Model of the World*, New York: Guilford.

Blunt, A. and Rose, G. (1994) 'Women's colonial and postcolonial geographies', in A. Blunt and G. Rose (eds) *Writing Women and Space: Colonial and Postcolonial Geographies*, New York and London: Guilford, 1–28.

Bommes, M. and Wright, P. (1982) 'Charms of residence: the public and the past', in R. Johnson *et al.* (eds) *Making Histories*, London: Hutchinson, 253–305.

Boniface, P. and Fowler, P. J. (1993) *Heritage and Tourism in the 'Global Village'*, London and New York: Routledge.

Bordo, S. (1986) 'The Cartesian masculinization of thought', *Signs* 11, 3: 439–456.

Boyer, M. C. (1992) 'Cities for sale: merchandising history at South Street Seaport', in M. Sorkin (ed.) *Variations on a Theme Park: The New American City and the End of Public Space*, New York: The Noonday Press, 181–204.

Brien, A. (1981) 'Alan Brien's London', *Punch*, 10 June: 6–7.

Brisbane City Council (1993a) *Livable Brisbane: A Plan for Australia's Most Livable City*, Brisbane: Brisbane City Council.

—— (1993b) *Aboriginal Art Trail: JC Slaughter Falls*, Brisbane: Community Arts Unit, Brisbane City Council.

—— (1994a) '$10m ecotourism centre for Brisbane', *Inside Brisbane*, Newsletter of the Office of Economic Development, 3, 3: 1–3.

—— (1994b) *Brisbane Ecotourism Centre: Invitations for the Expressions of Interest*, Brisbane: Office of Economic Development, Brisbane City Council.

Brokensha, P. and Guldberg, H. (1992) *Cultural Tourism in Australia*. Report commissioned by the Department of the Arts, Sport, the Environment and Territories. Canberra: Australian Government Printing Service.

Bropho, R. (1980) *Fringedweller*, Sydney: Alternative Publishing Co-operative Ltd.

—— (1992) No Title, in Nyungah People of the Swan River (eds) *The Brewery Picket*, Perth: Nyungah People of the Swan River. [Political broadsheet.]

—— and Ansara, M. (1989) *Always Was, Always Will Be*, Melbourne: Australian Film Institute. [Documentary film.]

Brownill, S. (1990) *Developing London's Docklands: Another Great Planning Disaster?*, London: Paul Chapman.

Budd, L. and Whimster, S. (eds) (1992) *Global Finance and Urban Living: A Study of Metropolitan Change*, London: Routledge.

Burgin, V. (1993) 'The city in pieces', *New Formations* 20: 33–45.

Butler, J. (1990a) 'Gender trouble, feminist theory and psychoanalytic discourse', in L. J. Nicholson (ed.) *Feminism/Postmodernism*, London and New York: Routledge, 324–340.

—— (1990b) *Gender Trouble: Feminism and the Subversion of Identity*, London and New York: Routledge.

—— (1992) 'Discussion on Aronowitz, S. "Reflections on identity"', *October* 61: 108–120.

Cain, E. (1991) 'To sacred origins – through symbol and story', in M. Fox, *Creation Spirituality and the Dreaming*, Newtown: Millennium Books, 73–86.

Cain, P. J. and Hopkins, A. G. (1993) *British Imperialism: Innovation and Expansion 1688–1914*, London: Longman.

Carter, E., Donald, J. and Squires, J. (eds) (1993) *Space and Place: Theories of Identity and Location*, London: Lawrence and Wishart.

Carter, P. (1987) *The Road to Botany Bay*, London: Faber and Faber.

Cassis, Y. (1988) 'Merchant bankers and City aristocracy', *The British Journal of Sociology* 39: 114–120.

Chambers, I. (1987) 'Maps for the metropolis: a possible guide to the present', *Cultural Studies* 1, 1: 1–21.

—— (1994a) *Migrancy, Culture, Identity*, London and New York: Routledge.

—— (1994b) 'Leaky habitats and broken grammar', in G. Robertson, M. Mash, L. Tickner, J. Bird, B. Curtis and T. Putnam (eds) *Travellers' Tales: Narratives of Home and Displacement*, London and New York: Routledge, 245–249.

Churches, S. (1990) 'Aboriginal people and government responsibility and accountability', *Aboriginal Law Bulletin* 2, 47: 6–7.

—— (1992) 'Aboriginal heritage in the wild west: Robert Bropho and the Swan Brewery site', *Aboriginal Law Bulletin* 2, 56: 9–13.

—— (1993) 'Letter to the editor', *Australian Law Journal* 67: 239.

City of London Police (1994) *Camerawatch: Closed Circuit Television Scheme: Code of Practice*, London: City of London Police.

City Recorder (11 October 1986) 'Spitalfields redevelopment'.

CityVision (1987) *New Directions for Central Perth*, Claremont, WA: CityVision.

Clarke, W. (1989) *The City of London Official Guide*, London: Hobson Publishing.

Clifford, J. (1988) *The Predicament of Culture*, Cambridge, MA, and London: Harvard University Press.

—— (1990) 'On collecting art and culture', in R. Ferguson, M. Gever, T. T. Minh-ha and C. West (eds) *Out There: Marginalization and Contemporary Cultures*, New York: Museum of Contemporary Art/Cambridge, MA: MIT Press, 141–169.

—— (1992) 'Traveling cultures', in L. Grossberg, C. Nelson and P. Treichler (eds) *Cultural Studies*, London and New York: Routledge, 96–111.

—— (1994) 'Diasporas', *Cultural Anthropology* 9, 3: 302–338.

Coakley, J. (1992) 'London as an international finance centre', in L. Budd and S. Whimster (eds) *Global Finance and Urban Living: A Study of Metropolitan Change*, London: Routledge, 52–72.

Cohen, A. P. (1985) *The Symbolic Construction of Community*, London: Tavistock.

Cohen, E. (1988) 'Authenticity and commoditization in tourism', *Annals of Tourism Research* 15, 3: 371–386.

Colls, R. and Dodd, P. (1986) 'Englishness and the political culture', in R. Colls and P. Dodd (eds) *Englishness: Politics and Culture 1880–1920* Beckenham: Croom Helm, 29–61.

Commonwealth of Australia (1993) *Mabo: the High Court Decision on Native Title*. Discussion Paper. Canberra: Australian Government Publishing Service.

Community Development Group (1989) *Planning Our Future*, Spitalfields: Community Development Trust.

Corner, J. and Harvey, S. (1991) 'Mediating tradition and modernity: the heritage/enterprise couplet', in J. Corner and S. Harvey (eds) *Enterprise and Heritage: Crosscurrents in National Culture*, London and New York: Routledge, 45–75.

Corporation of London (1984) *Draft Local Plan*, Guildhall, London: Department of Architecture and Planning, Corporation of London.

—— (1986) *City of London Local Plan*, Guildhall, London: Department of Architecture and Planning, Corporation of London.

—— (1987) *Inquiry into Objections to the City of London Local Plan*, Guildhall, London: Corporation of London.

—— (1993a) *Schedule of Development*, Guildhall, London: Department of Architecture and Planning, Corporation of London.

—— (1993b) *Security Cameras, Planning Advice Note 1*, Guildhall, London: Department of Planning and Building Security, Corporation of London.

—— (1993c) *The Way Ahead: Traffic and the Environment*, Draft Consultation Paper, Guildhall, London: Office of the Town Clerk, Corporation of London.

—— (1993d) 'Business as usual for the city', Press Release, 24 April.

—— (1993e) *New Traffic Arrangements for the City of London*, Guildhall, London: Department of Architecture and Planning, Corporation of London.

—— (1994) *The City of London to the Year 2000 and Beyond: Prospects for Office Demand*. Seminar proceedings, 16 May, London: The Royal Institute of Chartered Surveyors, Investment Property Forum and The Corporation of London.

Cosgrove, D. and Daniels, S. (eds) (1988) *The Iconography of Landscape*, Cambridge: Cambridge University Press.

Coupland, A. (1992) 'Docklands: dream or disaster?', in A. Thornley (ed.) *The Crisis of London*, London: Routledge, 149–162.

Craik, J. (1994) 'Peripheral pleasures: the peculiarities of post-colonial tourism', *Culture and Policy* 6: 153–182.

Crapanzano, V. (1992) 'The postmodern crisis: discourse, parody, memory', in G. E. Marcus (ed.) *Reading Cultural Anthropology*, Durham, NC and London: Duke University Press, 87–102.

Crawford, M. (1993) *One Money for Europe? The Economics and Politics of Maastricht*, London: Macmillan.

Cronon, W. (1991) *Nature's Metropolis: Chicago and the Great West*, New York: Norton.

Cruickshank, D. (1989) 'The past and MacCormac', *The Architectural Review* May: 68–75.

Crush, J. (1992) 'Beyond the frontier: the new South African historical geography', in C. Rogerson and J. McCarthy (eds) *Geography in a Changing South Africa*, Cape Town: Oxford University Press, 10–37.

—— (1994) Post-colonialism, de-colonization, and geography', in A. Godlewska and N. Smith (eds) *Geography and Empire*, Oxford: Blackwell, 333–350.

CSSD (1987) *Spitalfields Defender*, Broadsheet, London: CSSD.

—— (1988) *Petition to the House of Commons*, London: House of Commons.

—— (1989) *Petition to the House of Lords*, London: House of Lords.

Cullen, G. (1961) *Townscape*, London: The Architectural Press.

Culler, J. (1981) 'Semiotics of tourism', *American Journal of Semiotics* 1, 1–2: 127–140.

Daniels, S. (1989) 'Marxism, culture and the duplicity of landscape', in R. Peet and N. Thrift (eds) *New Models in Geography*, vol. II, London: Unwin Hyman, 196–220.

—— (1993) *Fields of Vision: Landscape Imagery and National Identity in England and the United States*, Cambridge: Polity Press.

Davies, H. (1992) 'Thoughts from a "dangerous man"', *The Spectator* 5 September, 12.

Davis, M. (1990) *City of Quartz*, London: Verso.

de Burgh, W. and de Burgh, M. (1981) *The Breakaways*, Perth: St Georges Books.

de Certeau, M. (1984) *The Practice of Everyday Life*, Berkeley: University of California Press.

—— (1986) *Heterologies: Discourse on the Other*, trans. Brian Massumi, Manchester: Manchester University Press.

de Cronin Hastings, H. (1944) 'Exterior furnishing or sharawaggi: the art of making urban landscape', *The Architectural Review*, January: 3–8.

—— (1945) 'Programme for the City of London', *The Architectural Review*, June: 158–187.

DEGW (1985) *Accommodating the Growing City*, Report by DEGW for Rosehaugh Stanhope Plc, London: DEGW.

Demeritt, D. (1994) 'The nature of metaphors in cultural geography and environmental history', *Progress in Human Geography* 18, 2: 163–185.

de Norbury Rogers, A. E. (1994) *Business Climate in Brisbane: A City of Growth*, Brisbane: Brisbane City Council.

Derrida, J. (1982) *Margins of Philosophy*, Chicago: University of Chicago Press.

—— (1992) *The Other Heading: Reflections on Today's Europe*, Bloomington and Indianapolis: Indiana University Press.

Deutsche, R. (1991) 'Boys town', *Environment and Planning D: Society and Space* 9: 5–30.

de Wolfe, I. [de Cronin Hastings, H.] (1949) 'Townscape', *The Architectural Review*, December: 355–362.

Douglas, M. (1984) *Purity and Danger: An Analysis of the Concepts of Pollution and Taboo* [first published 1966], London: Ark Paperbacks.

Drengson, A. R. (1989) *Beyond Environmental Crisis: From Technocrat to Planetary Person*, New York: Peter Lang.

Driver, F. (1992) 'Geography's empire: histories of geographical knowledge', *Environment and Planning D: Society and Space* 10: 23–40.

Duncan, J. (1990) *The City as Text: The Politics of Landscape Interpretation in the Kandyan Kingdom*, Cambridge: Cambridge University Press.

—— and Ley, D. (eds) (1993) *Place/culture/representation*, London and New York: Routledge.

During, S. (1987) 'Postmodernism or postcolonialism?', *Textual Practice* 1, 1: 32–47.

—— (1992) 'Postcolonialism and globalization', *Meanjin* 51, 2: 339–353.

Eckersley, R. (1992) *Environmentalism and Political Theory: Towards an Ecocentric Approach*, London: UCL Press.

Egloff, B. J. (1989) 'Old Swan Brewery Site, Perth: A Report to the Minister for Aboriginal Affairs on Information Received and Measures Taken Towards Conciliation', unpublished report, Canberra: Anutech.

Engels, F. (1971) *The Condition of the Working Class in England in 1844*, trans. W. O. Henderson and W. H. Chaloner, Oxford: Basil Blackwell.

—— (1977) *Origin of the Family, Private Property and the State*, Moscow: Progress Publishers.

English Heritage (1988) *Proof of Evidence: No. 1 Poultry Public Inquiry*, London: Department of Environment.

Erens, B. (1993) *The Residents of Bethnal Green: Their Characteristics and Views*. Report of a survey for the Bethnal Green City Challenge Company, London.

Esher, L. (1983) *A Broken Wave*, London: Pelican Books.

Evening Standard (17 July 1992) 'Desperate days down at Spitalfields'.

Fainstein, S. S. (1994) *The City Builders: Property, Politics and Planning in London and*

New York, Cambridge, MA and Oxford: Blackwell.

Fanon, F. (1968) *The Wretched of the Earth*, trans. Constance Farrington, New York: Grove.

Ferguson, K. (1993) *The Man Question: Visions of Subjectivity in Feminist Theory*, Berkeley and Los Angeles: University of California Press.

Ferrier, E. (1990) 'Mapping power: cartography and contemporary cultural theory', *Antithesis* 1: 35–48.

Fielder, J. (1991) 'Purity and pollution: Goonininup/The Old Swan Brewery', *Southern Review* 24, 1: 34–42.

Financial Times (26 April 1993) 'The bombing of the City'.

Fitzsimmons, M. (1989) 'The matter of nature', *Antipode* 21, 2: 106–120.

Forman, C. (1989) *Spitalfields: A Battle for Land*, London: Hilary Shipman.

Foucault, M. (1980) 'The eye of power', in C. Gordon (ed.) *Michel Foucault Power/Knowledge: Selected Interviews and Other Writings 1972–1977*, trans. C. Gordon, L. Marshall, J. Mepham and K. Soper, New York: Pantheon Books, 146–165.

—— (1986) 'Of other spaces', trans. J. Miskowiec, *diacritics* 16: 22–27.

Fourmile, H. (1989) 'Some background to issues concerning the appropriation of Aboriginal imagery', in S. Cramer (ed.) *Postmodernism: A Consideration of the Appropriation of Aboriginal Imagery*, Forum Papers, Brisbane: Institute of Modern Art.

Fox, M. (1991) 'Creation spirituality and the Dreaming', in M. Fox (ed.) *Creation Spirituality and The Dreaming*. Newtown: Millennium Books.

Fox, W. (1990) *Toward a Transpersonal Ecology: Developing New Foundations for Environmentalism*, Boston: Shambhala.

Frankenberg, R. and Mani, L. (1993) 'Crosscurrents, crosstalk: race, "postcoloniality" and the politics of location', *Cultural Studies* 7, 2: 292–310.

Freud, S. (1990 [1919]) 'The "Uncanny"', in The Penguin Freud Library, Vol. 14 *Art and Literature*, London: Penguin.

Friedmann, J. (1966) 'The World City hypothesis', *Development and Change* 17, 1: 69–83.

—— and Wolff, G. (1982) 'World city formation: an agenda for research and action', *International Journal for Urban and Regional Research* 6: 309–344.

Fringe Dwellers of the Swan Valley Inc. (1990) Press statement, 11 January.

—— (1992) *The Brewery Picket*. Publicity broadsheet, Perth.

Frow, J. (1991) 'Tourism and the semiotics of nostalgia', *October* 57: 123–151.

Fuss, D. (1989) *Essentially Speaking: Feminism, Nature and Difference*, London and New York: Routledge.

Gardiner, S. (1990) 'Palumbo's palaces', *Observer Magazine*, 1 July: 22–27.

Gates, H. L. Jr (1984) *Black Literature and Literary Theory*, London and New York: Methuen.

Gelder, K. (1993) 'The politics of the sacred', *World Literature Today* 67, 3: 499–504.

—— and Jacobs, J. M. (1995a) '"Talking out of place": authorizing the Aboriginal sacred in postcolonial Australia', *Cultural Studies* 9, 1: 150–160.

—— and Jacobs, J. M. (1995b) 'Uncanny Australia', *Ecumene* 2, 2: 173–185.

Georgian Group (1980–1988) *Records* of the Georgian Group, London.

Giddens, A. (1990) *The Consequences of Modernity*, Cambridge: Polity.

Gilroy, P. (1987) *There Ain't No Black in the Union Jack*, London: Hutchinson.

—— (1993) *The Black Atlantic: Modernity and Double Consciousness*, London: Verso.

—— (1994) 'Urban social movements, "race" and community', in P. Williams and L. Chrisman (eds) *Colonial Discourse and Post-Colonial Theory: A Reader*, New York: Columbia University Press, 404–420. (First published in 1987.)

Girouard, M., Cruickshank, D. and Samuel, R. (1989) *The Saving of Spitalfields*, Tenth Anniversary Publication, London: Spitalfields Historic Buildings Trust.

Giroux, H. (1992) 'Resisting difference: cultural studies and the discourse of critical pedagogy', in L. Grossberg, C. Nelson and P. Treichler (eds) *Cultural Studies*, New York and London: Routledge, 199–212.

Glass, R. and Frenkel, M. (1946) 'How they live at Bethnal Green', *Contact: Britain Between West and East*, London: Contact Publications Ltd, 12–15.

Goldberg, D. T. (1993) '"Polluting the body politic": racist discourse and urban location', in M. Cross and M. Keith (eds) *Racism, the City and the State*, London and New York: Routledge, 45–61.

Goodhart, C. and Grant, A. (eds) (1986) *Business of Banking*, London: Gower Press.

Gottdiener, M. (1986) 'Culture, ideology and the sign of the city', in M. Gottdiener and A. Ph. Lagopoulos (eds) *The City and the Sign*, New York: Columbia University Press, 202–218.

Goulbourne, H. (1991) *Ethnicity and Nationalism in Post-Imperial Britain*, Cambridge: Cambridge University Press.

—— (1993) 'Aspects of nationalism and Black identities in post-imperial Britain', in M. Cross and M. Keith (eds) *Racism, the City and the State*, London and New York: Routledge, 177–193.

Green, N. (1984) *Broken Spears: Aboriginals and Europeans in Southwest of Australia*, Perth: Focus Education Services.

Greenblatt, S. (1991) *Marvellous Possession: The Wonder of the New World*, Oxford: Clarendon Press.

Gregory, D. (1994) *Geographical Imaginations*, Cambridge, MA and Oxford: Blackwell.

Gregson, R. (n.d.) *Swan Brewery Redevelopment*. Publicity brochure, Perth: Roger Gregson Architects.

—— (1989) *Representation to an Inquiry under Section 10(4) of the Aboriginal and Torres Strait Islander Protection Act 1984, Concerning the Old Swan Brewery Site Perth, Western Australia*, Perth: Roger Gregson Architects.

Griffin, S. (1978) *Women and Nature: The Roaring Inside Her*, New York: Harper and Row.

Grossberg, L. (1988) 'Wandering audiences, nomadic critics', *Cultural Studies* 2, 3: 377–391.

Guardian (2 July 1987) 'Prince's East End safari'.

—— (5 June 1990) 'Battlecries in Banglatown'.

—— (26 April 1993) 'Foreign banks high on list of victims'.

Gupta, A. and Ferguson, J. (1992) 'Beyond "culture": space, identity and the politics

of difference', *Cultural Anthropology* 7, 1: 6–23.
Gyford, J. (1985) *The Politics of Local Socialism*, London: George Allen and Unwin.
Hackney Gazette (3 July 1987) 'Charles in cash boost pleas to city slickers'.
Hall, C. (1993) *White, Male and Middle Class: Explorations in Feminism and History*, Cambridge: Polity Press.
Hall, S. (1990) 'Cultural identity and diaspora', in J. Rutherford (ed.) *Identity: Community, Culture, Difference*, London: Lawrence and Wishart, 222–237.
—— (1991a) 'The local and the global: globalization and identity', in A. D. King (ed.) *Culture, Globalization and the World-System: Contemporary Conditions for the Representation of Identity*, Basingstoke: Macmillan in association with the Department of Art and Art History, State University of New York at Binghamton, 19–40.
—— (1991b) 'Old and new identities, old and new ethnicities', in A. D. King (ed.) *Culture, Globalization and the World-System: Contemporary Conditions for the Representation of Identity*, Basingstoke: Macmillan in association with the Department of Art and Art History, State University of New York at Binghamton, 41–68.
—— (1993) 'Culture, community, nation', *Cultural Studies* 7, 3: 349–363.
Handler, R. (1987) 'Heritage and hegemony: recent works on historic preservation and interpretation', *Anthropological Quarterly* 60: 137–141.
Hannerz, U. (1991) 'Scenarios for peripheral cultures', in A. D. King (ed.) *Culture, Globalization and the World-System: Contemporary Conditions for the Representation of Identity*, Basingstoke: Macmillan in association with the Department of Art and Art History, State University of New York at Binghamton, 107–128.
Haraway, D. (1989) *Primate Visions: Gender, Race and Nature in the World of Modern Science*, New York: Routledge.
Harley, J. B. (1988) 'Maps, knowledge and power', in D. Cosgrove and S. Daniels (eds) *The Iconography of Landscape*, Cambridge: Cambridge University Press, 277–312.
—— (1992) 'Deconstructing the map', in T. J. Barnes and J. S. Duncan (eds) *Writing Worlds: Discourse, Text and Metaphor in the Representation of Landscape*, London and New York: Routledge, 193–230.
Harris, J. and Thane, P. (1984) 'British and European bankers 1880–1914: an "aristocratic bourgeoisie"?', in P. Thane, G. Crossick and R. Floud (eds) *The Power of the Past: Essays for Eric Hobsbawm*, Cambridge: Cambridge University Press, 215–234.
Hartsock, N. (1990) 'Foucault on power; a theory for women?', in L. Nicholson (ed.) *Feminism/Postmodernism*, New York: Routledge, Chapman and Hall, 157–175.
Harvey, D. (1973) *Social Justice and the City*, London: Edward Arnold.
—— (1989) *The Condition of Postmodernity*, Oxford: Blackwell.
—— (1993) 'From place to space and back again: reflections on the condition of postmodernity', in J. Bird, B. Curtis, T. Putnam, G. Robertson and L. Tickner (eds) *Mapping the Futures: Local Cultures, Global Change*, London and New York: Routledge, 3–29.
Healey, P. (1990) 'Understanding land and property development processes: some key

issues', in P. Healey and R. Nabarro (eds) *Land and Property Development in a Changing Context*, London: Gower, 1–14.

Heritage Council of Western Australia (1991) *Swan Brewery, Perth: Conservation Analysis*. Draft report by Clive Lucas, Stapleto and Partners, Pty Ltd, Sydney. Perth: Heritage Council of Western Australia.

Hewison, R. (1987) *The Heritage Industry: Britain in a Climate of Decline*, London: Methuen.

Hobsbawm, E. and Ranger, T. (1983) *The Invention of Tradition*, Cambridge: Cambridge University Press.

hooks, b. (1991) *Yearning: Race, Gender and Cultural Politics*, London: Turnaround.

—— (1992) 'Representing whiteness in the black imagination', in L. Grossberg, C. Nelson and P. Treichler (eds) *Cultural Studies*, London and New York: Routledge, 338–346.

Howells, G. (1985) 'The politics of townscape: an appraisal of the postwar planning policy advanced by the "architectural review"', Unpublished dissertation for Diploma in Architecture, Plymouth Polytechnic.

HRH The Prince of Wales (1989) *A Vision of Britain: A Personal View of Architecture*, London: Doubleday.

Huggan, G. (1991) 'Decolonising the map: post-colonialism, post-structuralism and the cartographic connection', in I. Adam and H. Tiffin (eds) *Past the Last Post: Theorizing Post-Colonialism and Post-Modernism*, New York: Harvester-Wheatsheaf.

Huggins, J., Huggins, R. and Jacobs, J. M. (1995) 'Kooramindanjie: place and the postcolonial', *History Workshop Journal* 39, Spring: 165–181.

Hutcheon, L. (1994) 'The post always rings twice: the postmodern and the postcolonial', *Textual Practice* 8: 205–238.

—— (1995) 'Circling the downspout of empire: post-colonialism and postmodernism', in B. Ashcroft, G. Griffiths and H. Tiffin (eds) *The Post-colonial Studies Reader*, London and New York: Routledge, 130–135.

Independent, The, (22 September 1993) 'Dream scene'.

Jackson, P. (1985) 'Urban ethnography', *Progress in Human Geography* 9: 157–176.

—— (1991) 'Mapping meanings: a cultural critique of locality studies', *Environment and Planning A* 23: 215–228.

Jacobs, J. M. (1988) 'The construction of identity', in J. R. Beckett (ed.) *Past and Present: the Construction of Aboriginality*, Canberra: Aboriginal Studies Press.

—— (1992) 'Cultures of the past and urban transformation: the Spitalfields Market redevelopment in East London', in K. Anderson and F. Gale (eds) *Inventing Places*, Melbourne: Longman Cheshire, 194–214.

—— (1993) 'Shake 'im this country: the mapping of the Aboriginal sacred in Australia', in P. Jackson and J. Penrose (eds) *Constructions of Race, Place and Nation*, London: UCL Press, 110–118.

—— (1994a) 'The battle of Bank Junction: the contested iconography of capital', in S. Corbridge, R. Martin and N. Thrift (eds) *Money, Power, Space*, Oxford: Blackwell, 356–382.

—— (1994b) 'Earth honouring: western desires and indigenous knowledge', in A. Blunt and G. Rose *Writing Women and Space*, New York: Guilford, 169–196.

Jager, M. (1986) 'Class definition and the aesthetics of gentrification: Victoriana in Melbourne', in N. Smith and P. Williams (eds) *Gentrification of the City*, London: Unwin Hyman, 78–91.

Jameson, F. (1984) 'Postmodernism, or the cultural logic of late capitalism', *New Left Review* 146: 53–92.

—— (1986) 'Third world literature in the era of multinational capitalism', *Social Text* 15: 65–88.

—— (1991) *Postmodernism, or, The Cultural Logic of Late Capitalism*, Durham, NC: Duke University Press.

Jencks, C. (1988) *The Prince and the Architects and New Wave Monarchy*, London: Academy Editions.

Johnson, C. (1978) *Mies van der Rohe*, New York: Museum of Modern Art.

Johnston, V. (1988) 'Among others: reply to "Black Canberra"', *Art & Text* 30, September: 98–100.

Jones Lang and Wootton (1994) *Quarterly Review London City Offices, First Quarter*, London: Jones Lang and Wootton.

Karp, I. (1992) 'Introduction: Museums and communities: the politics of public culture', in I. Karp, C. Mullen Kreamer and S. D. Lavine (eds) *Museums and Communities: The Politics of Public Culture*, Washington and London: Smithsonian Institution Press, 1–17.

Kearns, G. and Philo, C. (eds) (1993) *Selling Places: The City as Cultural Capital, Past and Present*, Oxford: Pergamon.

Keith, M. and Cross, M. (1993) 'Racism and the postmodern city', in M. Cross and M. Keith (eds) *Racism, the City and the State*, London and New York: Routledge, 1–31.

Keith, M. and Pile, S. (1993a) 'The politics of place', in M. Keith and S. Pile (eds) *Place and the Politics of Identity*, London and New York: Routledge, 1–21.

—— (eds) (1993b) *Place and the Politics of Identity*, London and New York: Routledge.

King, A. D. (1976) *Colonial Urban Development: Culture, Social Power and Environment*, London: Routledge and Kegan Paul.

—— (1990a) *Urbanism, Colonialism and the World-Economy*, London and New York: Routledge.

—— (1990b) *Global Cities: Post-Imperialism and the Internationalisation of London*, London and New York: Routledge.

—— (1991a) 'Introduction: spaces of culture, spaces of knowledge', in A. D. King (ed.) *Culture, Globalization and the World-System: Contemporary Conditions for the Representation of Identity*, Basingstoke: Macmillan in association with the Department of Art and Art History, State University of New York at Binghamton, 1–18.

—— (1991b) 'The global, the urban and the world', in A. D. King (ed.) *Culture, Globalization and the World-System: Contemporary Conditions for the Representation of*

Identity, Basingstoke: Macmillan in association with the Department of Art and Art History, State University of New York at Binghamton, 149–154.

—— (1992) 'Rethinking colonialism: an epilogue', in N. Alsayyad (ed.) *Forms of Dominance: on the Architecture and Urbanism of the Colonial Enterprise*, Aldershot: Avebury, 339–355.

Kleinert, S. (1988) 'Black Canberra', *Art & Text* 30 June/August: 92–95.

Knudston, P. and Suzuki, D. (1992) *Wisdom of the Elders*, Toronto: Allen and Unwin.

Kong, L. (1993) 'Ideological hegemony and the political symbolism of religious buildings in Singapore', *Environment and Planning D: Society and Space* 11, 1: 23–46.

Kristeva, J. (1991) *Strangers to Ourselves*, New York: Columbia University Press.

—— (1993) *Nations without Nationalism*, trans. L. S. Roudiex, New York: Columbia University Press.

Kutcher, A. (1976) 'The views of St Paul's Cathedral', in D. Lloyd *et al.* (eds) *Save the City*, London: SPAB.

Langton, M. (1993) *'Well I heard it on the radio and saw it on the television . . .'*, North Sydney: Australian Film Commission.

Lash, S. (1990) *Sociology of Postmodernism*, London: Routledge.

—— and Urry, J. (1987) *The End of Organised Capitalism*, Cambridge: Polity Press.

—— and Urry, J. (1994) *Economies of Signs and Space*, London: Sage.

Lattas, A. (1990) 'Aborigines and contemporary Australian nationalism: primordiality and the cultural politics of otherness', in J. Marcus (ed.) *Writing Australian Culture: Text, Society and National Identity*. Special issue of *Social Analysis: Journal of Cultural and Social Practice* 27: 50–69.

Laurance, P. (1994) *Tourism in Brisbane: A City of Growth*, Brisbane: Brisbane City Council.

Lavarch, M. (1994) 'Foreword', in Attorney-General's Department *Native Title, Native Title Act 1993*, Canberra: Australian Government Press.

Lawlor, R. (1991) *Voices of the First Australians: Awakenings in the Aboriginal Dreamtime*. Rochester, VT: Inner Traditions.

Leech, K. (1994) *Brick Lane 1978: The Events and their Significance*, London: Stepney Books Publications.

Lefebvre, H. (1991) *The Production of Space*, trans. D. Nicholson-Smith (1974), Oxford: Blackwell.

Lisle-Williams, M. (1984) 'Merchant banking dynasties in the English class structure: ownership, solidarity and kinship in the City of London, 1850–1960', *The British Journal of Sociology* xxxv: 333–362.

Livingstone, D. N. (1992) *The Geographical Tradition*, Oxford and Cambridge, MA: Blackwell.

Lloyd, D. *et al.* (1976) *Save the City: A Conservation Study of the City of London*, London: Society for the Protection of Ancient Buildings.

London Borough of Tower Hamlets (1986) *Development Committee Report on Spitalfields Market*, London: LBTH.

London Edinburgh Trust and Grand Metropolitan (n.d.) *Building a Partnership from*

Brick Lane to Bishopsgate. Publicity brochure, London: LET and Grand Metropolitan.
London Planning Advisory Committee (1993) *The London Office Market: 1993 Update Report*, London: LPAC in conjunction with Investment Property Databank and Applied Property Research.
Lowenhaupt Tsing, A. (1994) 'From the margins', *Cultural Anthropology* 9, 3: 279–297.
Lowenthal, D. (1986) *The Past is a Foreign Country*, Cambridge: Cambridge University Press.
MacCannell, D. (1976) *The Tourist: A New Theory of the Leisure Class*, New York: Methuen.
McClintock, A. (1992) 'The angel of progress: pitfalls of the term "post-colonialism"', *Social Text* 10, 2 and 3: 84–98.
MacCormac, R. (1983) 'MacCormac's manifesto', *Architect's Journal* 15 June: 53–72.
McDowell, L. and Court, G. (1994) 'Performing work: bodily representations in merchant banks', *Environment and Planning D: Society and Space* 12: 727–750.
McLean, I. (1993) 'White Aborigines: cultural imperatives of Australian colonialism', *Third Text* 22, Spring: 17–26.
McRae, H. and Cairncross, F. (1985) *Capital City: London as a Financial Centre*, London: Methuen.
Maddock, K. (1987) 'Yet another "sacred site": the Bula controversy', in B. Wright, G. Fry and L. Petchkovsky (eds) *Contemporary Issues in Aboriginal Studies*, Sydney: Firebird Press.
—— (1988) 'God, Caesar and Mammon at Coronation Hill', *Oceania* 58: 305–408.
Mally, P. (1914) *Mount Coot-tha (One Tree Hill) Brisbane: An Amphitheatre of Unrivalled Grandeur*, Brisbane: Queensland Government Intelligence and Tourist Bureau.
Marcus, G. (1992) 'Past, present and emergent identities: requirements for ethnographies of late twentieth-century modernity worldwide', in S. Lash and J. Friedman (eds) *Modernity and Identity*, Oxford: Blackwell, 309–330.
Marcuse, H. (1964) *One-Dimensional Man: Studies in the Ideology of Advanced Industrial Society*, Boston, MA: Beacon Press.
Marks, S. (1984) *Corporation of London: Appeals by Number One Poultry Limited and City Acre Property Investment Trust Limited*. Report of the Inspector to the Secretary of State for the Environment, London: Department of Environment.
Marx, K. and Engels, F. (1976) *On Colonialism*, Moscow: Progress Publishers.
Massey, D. (1984) *Spatial Divisions of Labour: Social Structures and the Geography of Production*, London: Macmillan.
—— (1991) 'A global sense of place', *Marxism Today* June: 24–29.
—— (1992) 'Politics and space/time', *New Left Review* 196, November–December: 65–84.
—— (1993a) 'Power-geometry and a progressive sense of place', in J. Bird, B. Curtis, T. Putnam, G. Robertson and L. Tickner (eds) *Mapping the Futures: Local Cultures, Global Change*, London and New York: Routledge, 59–69.
—— (1993b) 'Politics and space/time', in M. Keith and S. Pile (eds) *Place and the*

Politics of Identity, London and New York: Routledge, 141–161.
—— (1994) *Space, Place and Gender*, Cambridge: Polity.
—— (1995) 'Places and their pasts', *History Workshop Journal* 39, Spring: 182–192.
Mathews, F. (1992) *The Ecological Self*, London: Routledge.
Matless, D. (1990) 'Ages of English design: preservation, modernism and tales of their history', *Journal of Design History* 3: 203–212.
Mercer, C. and Grundy, R. (1993) *Cultural Development in South East Queensland: A Policy Paper of the SEQ 2001 Project*, Brisbane: Regional Planning Advisory Group SEQ 2001.
Mercer, C. and Taylor, P. (1993) *A Cultural Development Strategy – Towards a Cultural Policy for Brisbane*, Brisbane: Institute for Cultural Policy Studies/Griffith University for the Brisbane City Council.
Merchant, C. (1980) *The Death of Nature: Women, Ecology, and the Scientific Revolution*, San Francisco: Harper and Row.
—— (1992) *Radical Ecology: The Search for a Livable World*, New York: Routledge.
Merlan, F. (1991) 'The limits of cultural constructionism: the case of Coronation Hill', *Oceania* 61: 341–352.
Meyrowitz, J. (1985) *No Sense of Place: The Impact of Electronic Media on Social Behaviour*, New York and Oxford: Oxford University Press.
Michaels, E. (1993) *Bad Aboriginal Art and Other Essays*, Minneapolis: University of Minnesota Press: and reproduced in Frow, J. and Morris, M. (eds) (1993) *Australian Cultural Studies: A Reader*, St Leonards: Allen and Unwin, 47–65.
Mickler, S. (1989) 'The Battle for Goonininup: A Charting of Post-colonial Representation in the City of Perth'. Unpublished manuscript.
—— (1991a) 'The Battle for Goonininup: Late Colonial Representation and the Old Brewery'. Unpublished manuscript.
—— (1991b) 'The battle for Goonininup', *Arena* 96: 69–88.
—— (1992) 'From the picket line', in Fringe Dwellers of the Swan Valley Inc. (eds) *The Brewery Picket*. Broadsheet 1.
Miles, R. (1982) *Racism and Migrant Labour*, London: Routledge and Kegan Paul.
Miller, H. (1966) *Tropic of Cancer*, St Albans: Panther.
Minh-ha, T. T. (1990) 'Cotton and iron', in R. Ferguson, M. Gever, T. T. Minh-ha and C. West (eds) *Out There: Marginalization and Contemporary Culture*, New York: Museum of Contemporary Art/Cambridge, MA and London: MIT Press, 327–336.
—— (1991) *When the Moon Waxes Red: Representation, Gender and Cultural Politics*, New York and London: Routledge.
—— (1994) 'Other than myself/my other self', in G. Robertson, M. Mash, L. Tickner, J. Bird, B. Curtis and T. Putnam (eds) *Travellers' Tales: Narratives of Home and Displacement*, New York and London: Routledge, 9–26.
Mishra, V. and Hodge, B. (1991) 'What is post-colonialism?', *Textual Practice* 5, 3: 399–413.
Mitchell, W. J. T. (ed.) (1994) *Landscape and Power*, Chicago and London: University of Chicago Press.

Mohanty, C. T. (1991a) 'Cartographies of struggle: Third World women and the politics of feminism', in C. T. Mohanty, A. Russo and L. Torres (eds) *Third World Women and the Politics of Feminism*, Bloomington and Indianapolis: Indiana University Press, 1–47.

—— (1991b) 'Under western eyes: feminist scholarship and colonial discourses', in C. T. Mohanty, A. Russo and L. Torres (eds) *Third World Women and the Politics of Feminism*, Bloomington and Indianapolis: Indiana University Press, 51–80.

Montgomery, J. and Thornley, A. (eds) (1990) *Radical Planning Initiatives: New Directions for Urban Planning in the 1990s*, London: Gower.

Morgan, G. (1993) 'Frustrated respectability: local culture and politics in London's Docklands', *Environment and Planning D: Society and Space* 11: 523–541.

Morris, M. (1988a) 'At Henry Parkes Motel', *Cultural Studies* 2, 1: 1–47.

—— (1988b) *The Pirate's Fiancée: Feminism Reading Postmodernism*, London: Verso.

—— (1990) 'Metamorphoses at Sydney Tower', *New Formations* 11: 5–18.

—— (1991) 'The man in the mirror: David Harvey's "Condition of Postmodernity"', in E. Jacka (ed.) *Continental Shift: Globalisation and Culture*, Sydney: Local Consumption Publications, 25–31.

Morris, W. (1986 [1890]) *News from Nowhere*, in A. L. Morton (ed.) *Three Works by William Morris*, London: Lawrence and Wishart.

Mowljarlai, D. and Peck, C. (1987) 'Ngarinyin cultural continuity: a project to teach young people the culture, including the re-painting of Wandjina rock art sites', *Australian Aboriginal Studies* 2: 71–78.

Muecke, S. (1992) *Textual Spaces: Aboriginality and Cultural Studies*, Sydney: University of New South Wales Press.

Mythen, F. P. (1991) 'The Spitalfields Development Group: A Universal Model for Community-based Response to Development Pressure?'. Unpublished M.Phil. thesis, Bartlett School of Architecture and Planning, University College London.

National Trust of Western Australia to Minister for Education and Planning (6 February 1986), reproduced in M. Ansara (1989) *Always Was, Always Will Be*, Sydney: Jequerity Pty Ltd.

National Trust of Western Australia (1987) 'Swan Brewery site'. Press release, 11 February.

Oldham, R. and Oldham, J. (1961) *Western Heritage: A Study of the Colonial Architecture of Perth, Western Australia*, Perth: Lamb Publications Pty Ltd.

Parry, B. (1987) Problems in current theories of colonial discourse', *Oxford Literary Review* 9, 1 and 2: 27–58.

Peach, C. (1986) 'Patterns of Afro-Caribbean migration and settlement in Great Britain: 1945–81', in C. Brock (ed.) *The Caribbean in Europe*, London: Frank Cass, 62–84.

Peachment, A. (1991) 'WA Inc', in A. Peachment (ed.) *The Business of Government: Western Australia 1983–1990*, Sydney: The Federation Press, 188–218.

Perth Sunday Times (9 August 1987) 'Wagyl eyes the brewery site'.

—— (5 August 1990) 'New brewery map upsets protesters'.

Plender, J. and Wallace, P. (1985) *The Square Mile: A Guide to the New City of London*, London: Century Publishing.

Pratt, G. (1992) 'Spatial metaphors and speaking positions', *Environment and Planning D: Society and Space* 10: 241–244.

Pratt, M. L. (1992) *Imperial Eyes: Travel Writing and Transculturation*, London and New York: Routledge.

Pred, A. (1986) *Place, Practice and Structure: Social and Spatial Transformation in Southern Sweden 1750–1850*, Cambridge: Cambridge University Press.

Probyn, E. (1990) 'Travels in the postmodern: making sense of the local', in L. J. Nicholson (ed.) *Feminism/Postmodernism*, London and New York: Routledge, 176–189.

Pryke, M. (1991) 'An international city going "global": spatial change and office provision in the City of London', *Environment and Planning D: Society and Space* 9: 197–222.

—— (1994) 'Looking back on the space of a boom: (re)developing spatial matrices in the City of London', *Environment and Planning A* 26, 2: 167–332.

Pudup, M. B. (1994) 'Trading places', *Antipode* 26, 2: 116–121.

Queensland Tourist and Travel Corporation (1992) *Vision 2000 – The Corporate Plan for the Queensland Tourism and Travel Corporation*, Brisbane: Queensland Tourist and Travel Corporation.

Raban, J. (1988) *Soft City*, London: Collins Harvill.

Rex, J. (1973) *Race, Colonialism and the City*, London: Routledge.

—— (1981) 'A working paradigm for race relations research', *Ethnic and Racial Studies* iv, 1: 1–24.

—— and Tomlinson, S. (1979) *Colonial Immigrants in a British City: A Class Analysis*, London: Routledge.

Richard Saunders and Partners (1986) *Office Floor Space Survey*, London: Richard Saunders and Partners.

Robertson, G., Mash, M., Tickner, L., Bird, J., Curtis, B. and Putnam, T. (eds) (1994) *Travellers' Tales: Narratives of Home and Displacement*, London and New York: Routledge.

Robins, K. (1991) 'Tradition and translation: national culture in its global context', in J. Corner and S. Harvey (eds) *Enterprise and Heritage: Cross-currents in National Culture*, New York and London: Routledge, 21–44.

Roddewig, R. (1978) *Green Bans: The Birth of Australian Environmental Politics*, Washington: Conservation Foundation/Sydney: Hale and Iremonger.

Roe, P. (1988) 'The children's country', *Meanjin* 47, 1: 11–20.

Rosaldo, R. (1989) 'Imperialist nostalgia', *Representations* 26: 107–122.

Rosehaugh Stanhope (1986) *Spitalfields Redevelopment*. Press release, 1 August, London: Records of the Spitalfields Historic Buildings Trust.

Ross, R. and Telkamp, G. . (eds) (1985) *Colonial Cities: Essays on Urbanism in a Colonial Context*, Dordrecht: Martinus Nijhoff.

Routt, W. D. (1994) 'Are you a fish? Are you a snake?: An obvious lecture and some notes on "The Last Wave"', *Continuum: The Australian Journal of Media and Culture*.

Special issue on *Critical Multiculturalism* 8, 2: 215–231.

Rowse, T. (1992) 'Hosts and guests at Uluru', *Meanjin* 51, 2: 265–276.

—— (1993) 'Mabo and moral anxiety', *Meanjin* 52, 2: 229–252.

Royal Commission on Local Government in Greater London (1962) *Written Evidence. PP 1164*, 1959–1960, xviii.

Sackett, L. (1991) 'Promoting primitivism: conservationist depictions of Aboriginal Australians', *The Australian Journal of Anthropology* 2, 2: 233–246.

Said, E. W. (1978) *Orientalism*, London: Routledge and Kegan Paul.

—— (1983) 'Traveling theory', in *The World, the Text and the Critic*, Cambridge, MA: Harvard University Press.

—— (1989) 'Representing the colonized: anthropology's interlocutors', *Critical Inquiry* 15 (Winter): 205–225.

—— (1990a) 'Narrative, geography and interpretation', *New Left Review* 180: 81–97.

—— (1990b) 'Yeats and decolonization', in T. Eagleton, F. Jameson and E. Said (eds) *Nationalism, Colonialism and Literature*, Minneapolis: University of Minnesota Press, 69–98.

—— (1993) *Culture and Imperialism*, London: Chatto and Windus.

—— (1995) 'Afterword to the 1995 printing', in *Orientalism*, London: Penguin, 329–354.

Salecl, R. (1993) 'National identity and socialist moral majority', in E. Carter, J. Donald and J. Squires (eds) *Space and Place: Theories of Identity and Location*, London: Lawrence and Wishart, 101–110.

Samuel, R. (1989) 'The pathos of conservation', in M. Girouard, D. Cruickshank and R. Samuel *The Saving of Spitalfields*, London: Spitalfields Historic Buildings Trust, 135–171.

Sassen, S. (1991) *The Global City: New York, London, Tokyo*, Princeton, NJ: Princeton University Press.

Schaffer, K. (1988) *Women and the Bush: Forces of Desire in the Australian Cultural Tradition*, Cambridge: Cambridge University Press.

Schwartz, B. (1991) 'Where horses shit and a hundred sparrows feed: Docklands and East London during the Thatcher years', in J. Corner and S. Harvey (eds) *Enterprise and Heritage: Crosscurrents of National Culture*, New York and London: Routledge, 76–92.

Senior, C. (1989) 'Old Swan Brewery Perth: A Report to the Minister for Aboriginal Affairs under Section 10 (4) of the Aboriginal and Torres Strait Islander Heritage Act 1984'. Unpublished report. Department of Aboriginal Affairs, Canberra.

Sennet, R. (1990) *The Conscience of the Eye: The Design and Social Lives of Cities*, New York: Knopf.

Shohat, E. (1991) 'Imagining *terra incognita*: the disciplinary gaze of empire', *Public Culture* 3: 41–70.

—— (1992) 'Notes on the postcolonial', *Social Text* 31/32, Spring: 99–113.

—— and Stam, R. (1994) *Unthinking Eurocentrism: Multiculturalism and the Media*, London and New York: Routledge.

Sibley, D. (1988) 'Survey 13: purification of space', *Environment and Planning D: Society and Space* 6, 4: 409–421.

Simon, D. (1984) 'Third world colonial cities in context', *Progress in Human Geography* 8, 4: 493–514.

Slaughter, J. C. (1964) *The Great Experiment: A Study of Local Government in Brisbane*, Brisbane: Brisbane City Council.

Slemon, S. (1991) 'Modernism's last post', in I. Adam and H. Tiffin (eds) *Past the Last Post: Theorizing Post-colonialism and Post-modernism*, Hemel Hempstead: Harvester-Wheatsheaf, 1–12.

—— (1994) 'The scramble for postcolonialism', in C. Tiffin and A. Lawson (eds) *De-Scribing Empire: Post-colonialism and Texturality*, London and New York: Routledge, 15–32.

Smith, A. M. (1994) *New Right Discourse on Race and Sexuality*, Cambridge: Cambridge University Press.

Smith, D. (ed.) (1991) *Apartheid City and Beyond*, London: Routledge.

Smith, M. P. and Feagin, J. R. (1987) *The Capitalist City*, Oxford: Blackwell.

Smith, N. (1984) *Uneven Development: Nature, Capital, and the Production of Space*, New York: Blackwell.

—— (1986) 'Gentrification, the frontier, and the restructuring of urban space', in N. Smith and P. Williams (eds) *Gentrification of the City*, London: Unwin Hyman, 15–34.

—— (1994) 'Geography, empire and social theory', *Progress in Human Geography* 18, 4: 491–500.

—— and Godlewska, A. (1994) 'Introduction', *Geography and Empire*, Oxford: Blackwell.

—— and Katz, C. (1993) 'Grounding metaphor: towards a spatialized politics', in M. Keith and S. Pile (eds) *Place and the Politics of Identity*, London and New York: Routledge, 67–83.

Smith, S. J. (1989) *The Politics of Race and Residence*, Cambridge: Polity Press.

—— (1993) 'Residential segregation and the politics of racialization', in M. Cross and M. Keith (eds) *Racism, the City and the State*, London and New York: Routledge, 128–143.

Soja, E. (1989) *Postmodern Geographies*, London: Verso.

Sorkin, M. (ed.) (1992) *Variations on a Theme Park: The New American City and the End of Public Space*, New York: The Noonday Press.

Sparke, M. (1994) 'White mythologies and anemic geographies: a review', *Environment and Planning D: Society and Space* 12, 1: 105–124.

Spitalfields Development Group (1987) *Spitalfields Market Development*. Promotional brochure, London: SDG.

—— (1988) *Spitalfields: A Continuing Story*. Promotional brochure, London: SDG.

—— (1991) 'Spitalfields Development Group launches new plans for Market site'. Press release, 15 October.

Spitalfields Heritage Centre (1987) *Heritage Centre for Spitalfields*. Promotional brochure, London: Spitalfields Heritage Centre.

Spitalfields Trust (1977) Untitled press release, London: Records of the Spitalfields Historic Buildings Trust.

—— (1986a) *Planning Gain: Issues for Discussion*, London: Records of the Spitalfields Historic Buildings Trust, June.

—— (1986b) *Spitalfields Trust, Director's Report*, London: Records of the Spitalfields Historic Buildings Trust.

—— (1989) *Spitalfields Trust: Tenth Anniversary*. Promotional brochure, London: Records of the Spitalfields Historic Buildings Trust.

—— (1992) *Assessment of Spitalfields Market Scheme*, London: Records of the Spitalfields Historic Buildings Trust.

Spitalfields Trust Newsletter (1976–1989) London: Records of the Spitalfields Historic Buildings Trust.

Spivak, G. C. (1988a) *In Other Worlds: Essays in Cultural Politics*, New York and London: Routledge.

—— (1988b) 'Can the subaltern speak?', in C. Nelson and L. Grossberg (eds) *Marxism and the Interpretation of Culture*, Urbana and Chicago: University of Illinois Press, 271–313.

—— (1990) *The Post-colonial Critic: Interviews, Strategies, Dialogues*, edited by S. Harasym, London and New York: Routledge.

—— and Young, R. (1991) 'Neocolonialism and the secret agent of knowledge', *The Oxford Literary Review* 12, 1–2: 220–251.

Stafford, L. (1992) 'London's financial markets: perspectives and prospects', in L. Budd and S. Whimster (eds) *Global Finance and Urban Living: A Study of Metropolitan Change*, London: Routledge, 31–51.

Stallybrass, P. and White, A. (1986) *The Politics and Poetics of Transgression*, London: Methuen.

Stamp, G. (1988) 'Britischer Architekt', *The Spectator* 1 October: 20–21.

Strawbridge, L. (1987/8) *Aboriginal Sites in the Perth Metropolitan Area: A Management Scheme*. Report for the Department of Aboriginal Sites, Western Australia Museum. Perth: Centre for Prehistory, University of Western Australia.

Sunday Telegraph (2 May 1993) 'NatWest Tower "should go"'.

Sunday Times (25 April 1993) 'City bloodied but unbowed'.

Survey of London (1957) *Spitalfields and Mile End New Town*, vol. XXVII, London: The Althone Press for London County Council.

Taussig, M. (1987) *Shamanism, Colonialism and the Wild Man*, Chicago and London: University of Chicago Press.

—— (1992) *The Nervous System*, London and New York: Routledge.

Taylor, P. (1993) *Political Geography: World-Economy, Nation-State and Locality*, 3rd edn, London: Longman.

Thomas, N. (1994) *Colonialism's Culture: Anthropology, Travel and Government*, Melbourne: Melbourne University Press.

Thompson, E. P. (1976) *William Morris: Romantic to Revolutionary*, New York: Pantheon Books.

Thrift, N. (1986) 'The geography of international economic disorder', in R. J. Johnson and P.

Taylor (eds) *A World in Crisis? Geographical Perspectives*, Oxford: Blackwell, 12–67.

—— (1989) 'Images of social change', in C. Hamnett, L. McDowell and P. Sarre (eds) *The Changing Social Structure*, London: Sage, 272–279.

—— (1994) 'On the social and cultural determinants of international financial centres: the case of the City of London', in S. Corbridge, R. Martin and N. Thrift (eds) *Money, Power and Space*, Oxford: Blackwell, 327–355.

—— and Williams, P. (1987) (eds) *Class and Space: The Making of Urban Society*, London: Routledge.

Tilbrook, L. (1985) *Report on the Aboriginal Significance of the Swan Brewery*. Report to the National Trust of Australia (WA). Perth: National Trust of Australia.

Timberlake, M. (ed.) (1985) *Urbanization in the World-Economy*, Orlando: Academic Press.

Times, The (27 April 1993) 'Company chiefs want steel gates over road'.

—— (1 February 1994) '£800m Rover sale to BMW sparks outcry'.

Todorov, T. (1984) *The Conquest of America: The Question of the Other*, trans. R. Howard, New York: Harper and Row.

Torgovnick, M. (1990) *Gone Primitive: Savage Intellects, Modern Lives*, Chicago: University of Chicago Press.

Toyne, P. and Johnston, R. (1991) 'Reconciliation, or the new dispossession?', *Habitat* 19: 3–10.

Traill, W. H. (1902) *Mt. Coot-tha Reserve: Description and Report*, Brisbane: Trustees of Mt Coot-tha.

Tugnutt, A. and Robertson, M. (1987) *Making Townscape*, London: Mitchell.

Urry, J. (1990) *The Tourist Gaze: Leisure and Travel in Contemporary Societies*, London: Sage.

—— (1995) *Consuming Places*, London and New York: Routledge.

Valuation Office, Inland Revenue (1988–1992) *Property Market Reports*, London: Inland Revenue.

Vidler, A. (1992) *The Architectural Uncanny: Essays in the Modern Unhomely*, Cambridge, MA: MIT Press.

Vinnicombe, P. (1989) *Goonininup: An Historical Perspective of Land Use and Associations in the Old Swan Brewery Area*, Perth: Department of Aboriginal Sites, Western Australian Museum.

von Sturmer, J. (1989) 'Aborigines, representation, necrophilia', *Art & Text* 32, Autumn: 127–139.

Wallerstein, I. (1974–1988) *The Modern World Systems*, 3 vols, New York: Academic Press.

—— (1979) *The Capitalist World-Economy*, Cambridge: Cambridge University Press.

Walton, J. (1976) 'Political economy of world urban systems: directions for comparative research', in J. Walton and L. Massotti (eds) *The City in Comparative Perspective*, London: Sage, 301–313.

Ware, V. (1992) *Beyond the Pale: White Women, Racism and History*, London: Verso.

Watts, M. J. (1991) 'Mapping meaning, denoting difference, imagining identity: dialectical images and postmodern geographies', *Geografiska Annaler* 73B: 7–16.

Weir, P. (1977) *The Last Wave*. Screenplay: Peter Weir, Tony Morphett and Petru Popesou. Director: Peter Weir.

Western Australia House of Assembly (1990) *Hansard*, 31 May: 1659.

Western Australia House of Representatives (1991) *Hansard*, 28 May: 10/41220.

West Australian, The (22 June 1988) 'Brewery tourist attraction'.

—— (22 June 1989) 'Myth is just tip of the iceberg claim'.

—— (26 June 1989) 'Police lashed over brewery swoop'.

—— (5 July 1989) 'Premier "not fit" to judge blacks'.

—— (10 July 1989) 'Tempers flare at old brewery protest'.

—— (30 September 1989) 'Brewery protestors to fight on'.

—— (21 June 1990) 'Bropho win throws laws into turmoil'.

—— (25 November 1990) 'Blacks vow to fight new brewery plan'.

Williams, P. and Chrisman, L. (1993) 'Colonial discourse and post-colonial theory: an introduction', in P. Williams and L. Chrisman (eds) *Colonial Discourse and Post-Colonial Theory: A Reader*, New York: Columbia University Press, 1–26.

Williams, R. (1977) 'The importance of community', in R. Williams *Resources of Hope*, London: Verso, 111–119.

—— (1980) 'Ideas of nature', in R. Williams *Problems of Materialism and Culture*, London: Verso, 67–85.

—— (1985) *The Country and the City*, London: The Hogarth Press. [First published 1973.]

Williams, S. (1992) 'The coming of the groundscrapers', in L. Budd and S. Whimster (eds) *Global Finance and Urban Living: A Study of Metropolitan Change*, London and New York: Routledge, 246–259.

Wilson, E. (1991) *The Sphinx in the City*, London: Virago.

Wolfe, P. (1994) 'Nation and MiscegeNation: discursive continuity in the post-Mabo era', *Social Analysis* 36, October: 93–151.

Wolff, J. (1991) 'The global and the specific: reconciling conflicting theories of culture', in A. D. King (ed.) *Culture, Globalization and the World-System: Contemporary Conditions for the Representation of Identity*, Basingstoke: Macmillan, in association with the Department of Art and Art History, State University of New York at Binghamton, 161–174.

—— (1993) 'On the road again: metaphors of travel in cultural criticism', *Cultural Studies* 7, 2: 224–239.

Woodward, R. (1991) 'Saving Spitalfields: The Politics of Opposition to Redevelopment in East London'. Unpublished Ph.D., QMWC, University of London.

—— (1993) 'One place, two stories: two interpretations of Spitalfields in the debate over its redevelopment', in G. Kearns and C. Philo (eds) *Selling Places: The City as Cultural Capital, Past and Present*, Oxford and New York: Pergamon, 253–266.

Worskett, R. (1988) *Proof of Evidence: No.1 Poultry Public Inquiry*. Public Inquiry statement, London: Department of Environment.

Wright, P. (1985) *On Living in an Old Country*, London: Verso.

—— (1987) 'The "heritage-thinking" that spells national decline', *The Listener* 24 September: 12–14.

—— (1991) *A Journey through Ruins: The Last Days of London*, London: Radius.

Yeo, S. (1986) 'Socialism, the state, and some oppositional Englishness', in R. Colls and P. Dodd (eds) *Englishness: Politics and Culture 1880–1920*, Beckenham: Croom Helm, 308–369.

Yeoh, B. (1991) 'Municipal Control, Asian Agency and the Urban Built Environment in Colonial Singapore, 1880–1929'. Unpublished D.Phil., Oxford University.

Young, I. M. (1990) *Justice and the Politics of Difference*, Princeton: Princeton University Press.

—— (1991) 'The ideal of community and the politics of difference', in L. J. Nicholson (ed.) *Feminism/Postmodernism*, London and New York: Routledge, 300–323.

Young, M. and Wilmott, P. (1957) *Family and Kinship in East London*, London, Routledge and Kegan Paul.

Young, R. (1990) *White Mythologies: Writing History and the West*, London and New York: Routledge.

—— (1995) *Colonial Desire: Hybridity in Theory, Culture and Race*, London: Routledge.

Žižek, S. (1989a) *The Sublime Object of Ideology*, London and New York: Verso.

—— (1989b) 'Looking awry', *October* 50: 31–56.

Zukin, S. (1986) 'Gentrification: culture and capital in the urban core', *Annual Review of Sociology* 13: 129–147.

—— (1988) *Loft Living: Culture and Capital in Urban Change*, London and New York: Radius.

—— (1991) *Landscapes of Power: From Detroit to Disney World*, Berkeley: University of California Press.

—— (1992) 'Postmodern urban landscapes: mapping culture and power', in S. Lash and J. Friedman (eds) *Modernity and Identity*, Oxford: Blackwell, 221–247.

INDEX

Aboriginality: and authenticity 134–5, 137, 147–9, 151; commodification of 134, 136, 161; constructions of 111–12, 123–4, 127, 161
Aborigines xii, 5, 6, 11, 18, 21, 23, 103–4; and architecture 121; and art production 144–53; and assimilation 108–9, 110; and authority 146–9; and colonialism 105–9; and environmentalism 136–8, 160; and koalas 132; and Nature 4, 11, 21, 32, 119, 121–4, 127, 131n, 132–9, 149, 155, 160; and Old Swan Brewery protest 110–15, 118, 124–7; as repressed Other 11, 105, 107–8, 126, 130, 137; spatial containment of 107–8; and tourism 132–6, 139, 140, 142–6, 150–3, 154–5, 160
Ackroyd, P. 70
Adam, I. 30
Africa 26, 27
Ahmad, A. 37n
Alsayyad, N. 20
Altab, Ali 91
Amery, C. 60
Anderson, K. 34, 37n, 99
Anderson, R. 48
Ansara, M. 124, 126, 127
anti-colonialism *see* postcolonialism
apartheid: 31
Appadurai, A. 5, 24, 33
Appiah, K. A. 26, 27, 155
appropriation 36, 68, 100, 160–1; of Aboriginality 123–4, 135–6, 142, 154; criticism of 143, 155; of heritage 58, 72, 87–90; of Nature 135–6
architecture 2, 20, 90–1; and Aborigines 121; and International Style 48–9, 59, 60; and Prince of Wales 41–3, 51, 69n, 86
Ashcroft, B. 25, 29, 37n

authenticity 36, 68, 80–1, 113, 133, 134–5, 137, 147–9, 151, 155
authority 8, 14, 22, 26, 28, 146–9, 152, 159, 162

Baker, L.M. 122, 137
Bangladesh 24, 96, 97, 101
Banglatown 99–101, 161
Bank Junction 38–64, 68, 68n, 85, 113; planning history of 43, 46–7; and townscape 47–51; *see also* No. 1 Poultry; Mansion House Square
Barnes, T. 9
Barthes, R. 4
Baudrillard, J. 154
Bauman, Z. 107
beer 110, 118; *see also* Old Swan Brewery; Truman's Brewery
Bell, Marshall 144, 145, 153
Ben Thompson Architects 90
Bengali Britons 10, 11, 70–102, 160, 162; racism towards 91–3; romantic views of 96–7
Benjamin, W. 147, 154
Berman, M. 74
Bhabha, H. 17, 25, 26–8, 40, 120, 130, 154
Birch, T. 122
Bishopsgate Goodsyard 72, 97–8
Blain, D. 76, 79, 80
Blaut, J. 17
Blunt, A. 7, 149
Bock, Dieter 63
Bommes, M. 35
Bond, Alan 109
Booth, Charles 74
Bordo, S. 6
Bourke, Brian 131n
Boyer, M.C. 36
Brewery Preservation Society 117–18
Brick Lane 91, 96

INDEX

Brisbane 5, 11, 133–56, 161, 162; Aboriginal remapping of 151–2; City Council 132–4, 143–5, 147–9; ecotourism centre 132–5, 146; *see also* J.C. Slaughter Falls; Mount Coot-tha
Bropho, R. 110–11, 121–2, 126, 127, 128
Butler, J. 8, 100, 148

Cain, E. 138
Campfire Consultancy 144
Canada 23
Canary Wharf 38, 56; *see also* Docklands
capitalism 6, 21, 23, 30; and culture 32–3, 35; and heritage 35, 72; late 30–4, 35
Carter, E. 40
Carter, P. 21–2, 105
cartography: and Aborigines 105, 107, 114–15, 119–21, 126–7; and imperialism 3, 6, 19–20, 21–2, 149–50, 158; and postcolonialism 121, 144, 149–54; and power 132
Chambers, I. 7, 34, 73, 146
Chrisman, L. 13, 16, 23, 28, 29
Churches, S. 108, 125, 126, 127
cities 9; and colonialism 4, 20–2, 105–9; as eroticised 115–16; as exoticised 31; First World 1, 4, 5, 6, 21, 31, 158; global 6, 10, 11, 24, 31, 38, 40, 41, 42, 53, 55, 67, 109; and imperialism 2, 4, 20–2, 74, 75; and Nature 11, 21, 122, 131n, 132–8; and otherness 4; as postcolonial 1, 4, 149; postmodern 1, 21, 31, 158; and race 4, 31–2, 72, 75; and sacred 112–13; and spectacle 4, 31; Third World 4, 6, 20, 158
City of London 5, 10, 11, 38–69, 72, 79, 83, 116, 159–60; financial restructuring of 38, 53–7, 60–1, 69n; foreign banks in 54, 64–5; as global city 38, 40, 41, 53, 55, 67, 109; and heritage 38, 40, 41, 46, 47, 49, 55–6, 60; and identity 39, 50, 53; and imperial nostalgia 50, 58; IRA bombings 64–7; office supply in 55–7, 69n; and otherness 41, 49, 58–2, 64, 68; and postimperialism 38, 40, 53, 58–9; relations with Germany 49–50, 58–64, 69n; sociology of 54, 68n
City Vision 116
Clifford, J. 5, 8, 11, 13, 23, 101
Coakley, J. 60
cognitive mapping 30, 34
Colls, R. 35

colonialism 15–19; ambivalence of 26–8, 107, 162; and Australia 18, 23–4; authority of 8, 26, 28, 107, 147; and cities 4, 20–2, 105–09; and culture 2, 14–15; definitions of 16, 37n; and India 18; and Nature 11, 21, 135–8, 140; settler-colonies 23; and space, x, 3–5
community: ideas of 74, 95–6, 101
Conrad, Joseph 51
Conservation Areas 43, 46, 47, 56, 68n, 76, 77, 89, 90
consumption 4, 31, 33, 35, 36, 104, 116, 145, 154, 158, 160
Corporation of London 39, 43, 46, 47, 50, 51, 55, 59, 65–7, 83
counter-colonial *see* postcolonialism
Court, G. 54
Craik, J. 136
Crapanzano, V. 29
Cronon, W. 122, 135
Cross, M. 31, 32, 72
Cruickshank, D. 60, 78, 90
Crush, J. 31
Cullen, G. 48, 51
culture: and capitalism 32–3; and colonialism 2, 14–15; and imperialism 2, 13, 14, 17

Daniels, S. 43, 50, 51, 113
Davis, M. 67
de Certeau, M. 10, 21, 107, 139, 149, 151
de Cronin Hastings, Hubert 48, 68n
Deep Ecology 136
Demeritt, D. 10
Derrida, J. 14, 41, 58
Deutsche, R. 30
diaspora xi, 4, 6, 13, 22, 24, 25, 26, 34, 36, 70, 97, 101–2, 161
difference 1, 2, 4, 7, 8, 26, 30, 32, 33, 41, 116, 148, 158, 162
discourse x, 2, 3, 8, 9, 13, 14, 28, 39
Docklands (London) 38, 55, 56; *see also* Canary Wharf; London
Dodd, P. 35
Douglas, M. 127
dreaming 103, 116, 130n
Driver, F. 6
Duncan, J. 9
During, S. 25
dwelling 9, 70, 102, 127, 138

East End (London) 73; as other 70, 74; as community 74; as racialised 74–5, 91; *see also* Spitalfields
ecocentrism 137

Empire 39–40, 47, 50, 53; idea of 51, 67–8
Engels, Friedrich 73
English Heritage 47
Englishness 5, 43, 48–9; 59, 71, 73, 74, 81; and Left 92, 101; and multiculturalism 86–7, 91, 96, 101–2
environmentalism: and Aborigines 136–8, 160
Esher, L. 49, 60
essentialism 2, 8–9, 28, 35, 36, 40, 111, 143, 148, 161, 162, 163; strategic 35, 100, 143, 148, 156
Eurocentrism 17, 27
Europe 6, 13, 41, 48–9, 55–6, 58–64

Fainstein, S.S. 98
Fanon, F. 17
feminism 3, 6, 7, 8, 29, 30, 34; ecofeminism 136
Ferguson, K. 8
Fielder, J. 122, 127
First World 13, 23, 33, 34; cities 1, 4, 5, 6, 21, 31, 34, 158
Fitzsimmons, M. 122, 135
Forman, C. 96, 97
fortress city 67
Foucault, M. 119
Fourth World 6, 23, 33, 34
Fox, M. 138
Frankenberg, R. 6, 25
Frankfurt 56, 60
Friedmann, J. 20, 31
fringedwellers 108, 110–11, 121, 126–7
Frow, J. 134

Gates, H.L. Jr. 154
gaze: imperial 139–41, 149, 159; of other 143; panoptic 37n, 139–41, 145; tourist 118–19
Gelder, K. 12n, 14, 115
gender 2, 3, 105, 137
gentrification 4, 11, 31, 35, 38, 72, 75–87, 143, 158, 160
geographies: imaginative 3; pragmatic x, 6; 'real' x, 3–5, 158; postcolonial 163; promiscuous 5; textual 6, 9–10; uneven 14, 16, 158; *see also* space
Georgian Group 85, 102n
Germany 49–50, 59-64, 69n
Gilbert and George 82
Gilroy, P. 32, 72, 120, 151
Girouard, M. 77
global 6, 9, 11, 24, 30, 38, 40, 41, 53, 55, 67, 104, 109, 116, 118, 149, 155
globalisation 31–4, 35, 36, 37n, 118

Godlewska, A. 3
Goldberg, D.T. 73, 75
Goonininup 107, 120
Gottdiener, M. 9
Greenblatt, S. 141
Gregory, D. 10, 73
Gregson, R. 120, 123, 131n
grids: and colonial urban development 21–2, 105–7
Griffiths, G. 37n
Grossberg, L. 6

Hall, C. 3
Hall, S. 71
Haraway, D. 136
Harley, J.B. 3, 19, 132, 149
Hartsock, N. 6
Harvey, D. 29, 30–1, 36, 131n
Hawksmoor, Nicholas 39, 46
Healey, P. 57
heritage 4, 11, 21, 34–6, 74, 102n, 154, 159; in City of London 38, 40, 41, 46, 47, 49, 55–6, 60; in Spitalfields 72, 74, 75–90; in Perth 115–19, 124, 131n; and rock art 146–7
Hewison, R. 35
Hobsbawm, E. 35
Hodge, B. 24
Holford, Sir William 116
hooks, b. xi, 7, 35, 111
Howells, G. 49
Huggan, G. 19
Huggins, J. 12n
Huggins, R. 12n
Huguenots 70, 75, 76, 81, 87, 101
Hutcheon, L. 25, 29
hybridity 4, 8, 13, 27–8, 30, 144, 150, 151, 152, 153, 162

identity: and beer 118; and imperialism 2; and other 40, 58; and performance 148; and place 1, 2, 3, 8, 10, 24, 33, 34–6, 37n, 40, 49–53, 73, 101–2, 118, 153, 156, 162; and postcolonialism 28, 29, 143, 155; and space 1, 4–5, 8
imperialism xi, 15–19; and anxiety 3, 22, 26, 159; and Australia 18; British 10; and cities 2, 4, 20–2, 74, 75; and culture 2, 13, 14, 17; definitions of 16, 37n; domesticated 58; idea of 51; and India 18; and the local 6, 72; lingering 1, 16; and nostalgia 4, 15, 21, 22, 34, 40, 50, 58, 68, 159, 163; and space 1, 3–5, 13, 19–22, 149–50; and urbanisation 4, 20–1, 73–5, 79, 159; variability of 17–19, 159;

vulnerability of 14, 107, 162
indigenous claims 23–4, 34–35, 107; *see also* land rights
IRA: 64–7

J.C. Slaughter Falls 143; art trail at 144–54.
Jackson, P. 36
Jacobs, J.M. 12n, 14, 111, 114, 137
Jager, M. 35
Jameson, F. 30, 32, 33, 34, 37n
Jencks, C. 60
Johnston, V. 146

Karp, I. 35
Katz, C. 3
Kearns, G. 33
Keith, M. 31, 32, 35, 37n, 38, 72
King, A.D. 16, 20, 21, 37n, 40, 53, 54
Kleinert, S. 146
Knudston, P. 136, 137
Kong, L. 113
Krier, Leon 86
Kristeva, J. 130, 163
Kutcher, A. 50

land rights xi, 4, 22, 23, 34, 35, 36, 100; anxieties over 112, 115, 161; in Australia 111–12, 122, 127, 147, 154; primitivism of 111–12, 154; urban 111–12;
landscape: as text 9–10
Langton, M. 142, 143
Lash, S. 32
Lattas, A. 137, 138
Lefebvre, H. 127
Left: and anti-racism 93, 162; and Bengali Spitalfields 93–7; and Englishness 92; and Spitalfields Market 93–6; and Thatcherism 92–3, 97
Ley, D. 9
Livingstone, D. 3, 149
Lloyd, D. 42, 47
local 6, 9, 11, 17, 33, 34, 35–6, 40, 58, 72, 104, 107, 155, 158
London 5, 10, 11, 15, 24, 31, 161; *see also* City of London; Docklands; East End; Spitalfields
Los Angeles 21, 31, 67
Lowenhaupt Tsing, A. 6
Lowenthal, D. 48
Luftwaffe 42, 60
Lund, Neils M. 38, 39, 53

Maastricht Treaty 58
Mabo decision 112; claims 154
McClintock, A. 23, 25, 26
MacCormac, Richard 89–90
McDowell, L. 54
McLean, I. x, 103
Maddock, K. 114, 119
Mally, P. 139, 140, 141
Mani, L. 6, 25
Manichaean binary 2, 103
Mansion House Square 42, 43, 46, 59; and Prince of Wales 59-60; *see also* Bank Junction
Marcus, G. 36
Marcuse, Herbert 73
Market: as other 85, 90 ; as split subject 84, 85
Marx, Karl 135
masculism 3, 6, 7
Massey, D. 3, 5, 6, 33, 34, 35, 36, 163
Mathews, F. 137
Matless, D. 49
Mayhew, Henry 74
mega-development 4, 11, 38, 72, 75, 91
Mercer, C. 145
Merchant, C. 136
Merlan, F. 121
Meyrowitz, J. 37n
Michaels, E. 147, 148, 149
Mickler, S. 108, 116, 119, 130n
Mies van der Rohe, Ludwig 42, 43, 46, 56, 57, 59–60
migration 24, 31, 34, 70, 71, 72, 74, 96, 97, 159
Miles, R. 31
Miller, H. 103
mimesis 26–28, 29, 30, 100, 105, 107, 157
Minh-ha T.T. 1, 7
Mishra, V. 24
Mohanty, C.T. 28–9, 30, 33–4
Morris, M. 6, 7, 10, 30, 134, 142, 149
Morris, William 21, 37n, 74, 76, 80
Mount Coot-tha 138; and imperial touring 138–142; and indigenous touring 142–154; and panoptic gaze 139–41, 145
Mount Eliza Depot 107–8
Muecke, S. 105
multiculturalism: in Australia 23, 116; in Britain 11, 73, 86–7, 96, 100–2, 160
museums 133, 136, 145; in Spitalfields 81–2, 87; in Western Australia 114, 115, 123–124, 131n
Mutitjulu Community 122, 137

National Trust of Western Australia 117
nationalism xi, 4, 24, 25, 28, 34, 35, 36, 37n, 91, 153, 163

Native Title 112, 131n
Nature: and Aboriginality 4, 11, 21, 32, 119, 121–4, 127, 131n, 132–9, 149, 160; and colonialism 135–8; and culture 135–6, 140, 160; feminised 3, 137; and modernity 136; and tourism 132–9, 154, 160
neo-colonialism 16, 24, 25, 31, 160; and environmentalism 136–8, 160; and tourism 123, 136, 142, 154–5, 160; variability of 155
New South Wales 18
New York 31, 56, 115, 140
Ngarinyin 146–7
Nilsen, Laurie 144, 145, 151, 152
No. 1 Poultry 42, 43, 46, 49, 50–1, 57–8, 60, 61–3; and German co-investor 63–4; planning history of 38, 39, 43, 46, 61–3; and Prince of Wales 51, 63–4; *see also* Bank Junction
nomadism 107, 110–11
Northern Ireland 64, 76
nostalgia: for authenticity 155; imperial 4, 15, 21, 22, 25, 33, 34, 40, 41, 43, 50, 58, 68, 159, 163; for primitive 4, 142, 161; for revolution 17, 33

Old Swan Brewery: and Aborigines 110, 115, 118, 119–29; as heritage 116–19, 121, 131n; and Nature 104, 119, 121–4, 127, 131n; redevelopment of 103–4, 107, 109–10; *see also* Perth; Waugal
orienteering 152–3, 154
Other 2, 4, 5, 8, 11, 13, 14, 25, 26, 27, 30, 33, 143, 158, 159, 162; Aborigines as 103, 107, 127, 130, 137, 151, 154, 158; and City of London 40, 41, 58–62, 64, 68; as Self 138; speaking for 8; and Spitalfields 70, 71, 72, 73, 74
otherness x, xiii, 4, 17, 25, 30, 34, 87, 102, 131n, 156, 160, 162, 163

Palumbo, Peter 42, 43, 46, 47, 56, 57–8, 59, 60, 62, 63, 68n
Parry, B. 28
pedagogy: postcolonial 152–3, 161
Perth 5, 11, 103–31, 159, 160, 161; and Aborigines 103, 104, 107–9, 110–11, 114, 119–21, 123, 126–7, 130, 131n, 162; and colonialism 105–9; as eroticised 115–16; and heritage 115–19; and multiculturalism 116; and restructuring 109, 116–19; and the Waugal 103, 107, 115, 117, 119–21, 123, 126–7, 130, 131n, 162;

see also Old Swan Brewery
Pevsner, Nikolaus 68
Philo, C. 33
Pile, S. 35, 37n, 38
place-making 40, 145–7, 162
place: cultural politics of 9; and heritage 34–6, 48, 72; and imperialism x, 35, 72; and identity 1, 2, 3, 8, 10, 24, 33, 34–6, 37n, 40, 49–53, 73, 101–2, 118, 153, 156, 162; no sense of 37n; progressive sense of 163
planning 2, 9, 10, 20, 21; and Brisbane 143–5; and City of London 38, 39, 41–3, 46, 47–9, 50, 52, 53, 59–64, 65–7; and colonisation 105; cultural 143, 144–5; and Perth 105, 109, 116–17, 119, 120, 127; and sacred 112–13, 114, 115; and Spitalfields 90, 93–5, 98–9, 102n
Plender, J. 53
Port Phillip 18
position: of author, x, 5, 12n, 15-16, 37n; speaking 7–10
postcolonialism xi, 2, 8, 26, 33, 34–5, 142; and agency 27–8, 148, 162–3; in Australia 23–4; and authenticity 146–7, 149; and authority 8, 14, 26, 28, 146–9, 152; in Britain 71, 97, 157; as complicit 24; as critique 13–14, 26–9; definitions and limits of 14–15, 22–9, 37n, 142, 161–2; and globalisation 25, 27, 35, 36; and nostalgia 25; and postmodernity 29–34; and space 1, 4, 34, 150–4, 161–2; and tourism 142–54, and tradition 146–9, 153, 162
postimperialism 38, 40, 41, 53, 58–9
postmodernity: and cities 1, 4, 29, 31–3, 131n, 158; and postcolonialism 29–34
poststructuralism 29–30
Powell, Enoch 71, 74
Pratt, M.L. 139, 140, 141
Pred, A. 39
primitivism 4, 103, 133–5, 136, 137–8, 142, 143, 147, 155, 156, 161
Prince of Wales 41–3, 51, 59–60, 63–4, 69n, 70, 86, 98, 99
Prince, H. 48
Probyn, E. 35
Pryke, M. 39, 55, 56, 57
Pudup, M.B. 122

Queensland 1, 12n, 132, 144

Raban, J. 74
race xii; as construct 2, 3, 17, 31,

INDEX

99–100; and cities 4, 24, 31–4, 72, 75, 116
racism x, xii,17, 24, 31, 71–2, 74; in Spitalfields 72, 91–3; in Perth 108, 111, 118
Ranger, T. 35
reconciliation 2, 11, 123, 124, 144, 161, 162; and morality 134–5
resistance 15, 21, 28, 36, 68, 100, 104, 119, 129, 147
Rex, J. 20, 31
Robertson, G. 7
Robins, K. 40, 71
Roe, P. 152
Rose, G. 7, 149
Rosehaugh Stanhope 86
Ross, R. 20
Rover cars 62–3
Rowse, T. 122, 135

Sackett, L. 136
sacred sites: as 'hidden' 113–14, 119; mapping of 114–15, 119–21, 126–7; and modernity 112–14, 121; non-disclosure of 113–14, 131n; protection of 111–12, as smudges 114
sacred: Aboriginal 11, 112; in cities 11, 112–13, 119; *see also* Waugal
Said, E. 3, 4, 5, 8, 9, 13, 16, 17, 19, 27, 37n, 150, 153
St Paul's Cathedral 39, 49–51, 53, 57
Salecl, R. 40
Samuel, R. 35, 39, 78, 80–1
Sassen, S. 31, 40, 41
SAVE Britain's Heritage 102n
Schaffer, K. 137
Schwartz, B. 75
secular: in cities 11, 113, 162
Self 2, 3, 5, 8, 11, 13, 14, 26, 68, 70, 71, 72, 73, 107, 127, 130, 151, 153, 159, 160, 162
Sennett, R. 112–13
Severs, Dennis 89
sharawaggi 48, 68n; *see also* townscape
Shohat, E. 3, 17, 37n
Sibley, D. 87
Simon, D. 20
Singapore 21, 113
Slemon, S. 27
Smith, A.M. 71
Smith, D. 31
Smith, N. 3, 75, 135
Smith, S.J. 32
social polarisation 32
Soja, E. 21, 29, 131n
Sorkin, M. 33
South Africa 31

sovereign: rights of 124–6
space: and imperialism 19–22; in-between 8; metaphorical x, 3; and postcolonial 1, 4, 33–4, 150–4, 161–2; and postmodern 31, 33–4; production of 9; as 'real' 3–5, 158; third 154; *see also* geographies
Spitalfields 5, 10–11, 24, 70–102, 160; and Banglatown 99–100; and Bengali community 11, 70, 81, 83, 85–7, 91–3, 95–7; community development of 70, 89, 98–9, 102n; gentrification 72, 75–87; Heritage Centre 87; and housing crisis 70, 97, 102n; and identity 72–3, 81, 95–6; and Market redevelopment 72, 83–91; multiculturalism 73, 86–7, 95–6, 100–2; neo-Georgian residents 80–2; opposition to Market redevelopment 93–6; as other 70, 71; politics of 92–3; and Prince of Wales 70, 86, 98; property prices 82, 83; racism 91–3; romantic views of 95–6; Survey of London 75–6; as unoccupied 79, 81
Spitalfields Development Group 83, 84, 87, 88, 89, 90, 91, 102n; and heritage 84, 87–90; and planning gains 89, 102n
Spitalfields Historic Building Trust 76–8; and Bengalis 81, 83, 85–7; as gentrifiers 79–85; and Market 77, 84–5, 87, 89, 90; and multiculturalism 87
Spivak, G.C. 3, 8, 30, 148
Stafford, L. 53
Stallybrass, P. 84
Stam, R. 17, 37n
Stamp, G. 63
Stevas, Lord St John 90
Stirling, James 42, 43, 46, 57, 61, 62–3
Strawbridge, L. 114
Stuttgart 63; Neue Staatsgalerie 62, 63
Survey of London 75–6
Suzuki, D. 136, 137
Swanke Hayden Connell 90
Sylhet 24

Taussig, M. 114
Taylor, P. 18
Telkamp, G. 20
Terra nullius 18, 105, 112, 159
Terry, Quinlan 86
Thatcherism: and inner city 74–5, 92
The Heart of the Empire 38, 39, 53
The Last Wave 103–4
Third World 4, 5, 6, 13, 20, 21, 23, 31, 33, 34, 70, 158; literature 37n; and

women 28
Thomas, N. 2, 14, 18, 19, 27, 137
Thompson, E.P. 74, 76
Thrift, N. 35, 54
Tiffin, H. 30, 37n
Tjakamarra, Michael Nelson 146
Tokyo 31, 56, 95, 115
Tomlinson, S. 31
Thompson, Ben 90
Torgovnick, M. 137
tourism 31; and Aboriginality 123–4, 132–6, 139, 140, 142–6, 150–3, 154–5, 160; cultural 132, 133, 135; and dwelling 138, 142; ecotourism 132, 133–5, 146; imperial 138–42; indigenous 142–54; as neo-colonialism 136; and Old Swan Brewery 103, 109, 116, 118–19, 123–4; as progress 142, 149; in Queensland 132–56; and Spitalfields 99
Tower Hamlets 70, 83–4
townscape: and City of London 41–2, 47–53; and hierarchy 52–3; and Old Swan Brewery 118–19; in Spitalfields 89; as visual policy 47–8, 68n
tradition 4, 134, 144, 146–9, 151, 153, 162
Traill, W.H. 139, 140, 141
travel 1, 6, 7; and feminist critique 6–7; writing 140
Truman's Brewery 72, 97–8

uncanny 28, 130
urbanisation 2, 4, 20–21, 73–5, 158, 159; and nihilism 73–4, 95
Urry, J. 32, 33, 35, 118

Victorian Society 102n

Vidler, A. 3, 127
Vinnicombe, P. 107, 108, 114, 121

WA Inc. 130–1n
Wadjina 146–7
Wallace, P. 53
Wallam, Reggie 108
Wallerstein, I. 16, 20
Ware, V. 3
Watts, M.J. 33
Waugal 103, 107, 115, 117, 130n, 131n, 159, 162; legal influence of 125–6; mapping of 119–21, 126–7; and nation 130; as polychrome path 123
Weir, P. 103
Western Australia Development Corporation 103–4, 109–10, 123, 124
Westlake, John 17
White, A. 84
Wilford, Michael 42, 43
Williams, P. 13, 16, 23, 28, 29
Williams, R. 74, 122, 135
Wilmott, P. 74
Wilson, E. 38, 39
Wolff, J. 7
world-system theory 4, 16, 17, 20, 21, 37n
Worskett, R. 47, 49, 51
Wright, P. 35, 39, 43, 69n, 75, 86

xenophobia 24, 40

Yeo, S. 92
Yeoh, B. 20, 21
Young, I.M. 36, 73, 87, 95, 101, 116
Young, M. 74
Young, R. 8, 17, 22, 27, 28, 30, 33

Žižek, S. 143
Zukin, S. 35–6, 37n